Old Souls On Resta

By Joan A. Aebig

DEDICATION

To my family who loves science fiction and joins me on my faith journey.

Table Of Contents

Cast Of Characters

Name	Home Planet	Occupation	Aura	Gender
Sarah	Earth		purple	she
Javi	Algenia	Agriculture	blue→gold	ke
Aron	Lanet	Miller	green	he
Voder	Frond	Conductor	black line	he
Zac	Frond	Soldier	green	he
Triol	Algenia	Golfer	blue	ke
Boulder	Far	Tech worker	blue	he
Pastor	Far	Pastor	purple	she
Caveman	Far	Tech worker	yellow	she
Bull	Far	Tech head	red	he
Dora	Narth	Miller		she

Chapter 1

Sarah And Meditation

Sarah hopped in her Toyota, picked up Karen and headed for church, the Methodist church in town. They were taking a month-long world religion class which turned out to be more difficult than they thought it would be.

"I really am surprised there are so many religions," said Sarah. "4300? Doesn't seem possible."

"I know", said Karen. "What religions are we covering in today's class?"

"I think we are learning about the ones that use meditation as a means of worship."

In the class, they learned that the mind focuses on a particular thought, activity, or object so that the person obtains a clear, calm state. Thus, the benefits are to reduce stress and anxiety and enhance self-awareness. The major religions use meditation and many of the minor ones do, too. The Hindus and Buddhists meditate the most. The class went on and on with karma, yoga, sutra, and Salah. They learned there are six types of meditation, e.g., mindfulness, transcendental, focused, etc.

By the end of the class, Sarah rolled her eyes.

"This is too much," she said. "I'm exhausted."

Karen nodded, "Yup."

That night, Sarah gave meditation a lot of thought. If millions of people from all different religions practice it, there must be something to it. In bed, Sarah closed her eyes and selected peace to focus on. But her mind wandered, and she found it difficult to stay on track. She tried to continue focusing on peace but ended up thinking about her soul. She hadn't thought about her soul very much. Everyone knew they had one, but they didn't think about their body and soul functioning together as one unit. The soul rarely poked its ephemeral arm outward, a reminder saying, "Here I am. See me." No one thought about that in daily life…until meditation revealed it.

****

Sarah tore through space. Or was it time? She heard screams. As the flight slowed she felt comfortable. A body presented itself and she entered it. But where were the screams coming from?

It's me! I'm screaming! she realized. *What's happening?*

"It's okay," someone said. "You're fine. You've made it."

3

Sarah easily calmed down, glad the dream had ended. As she became more awake she looked up to see a man looking down at her.

"Where am I?" she asked.

"You are on Planet Resta in the town of Sedona," he said.

"What?" she asked.

"You are on Planet Resta in the town of Sedona," he repeated.

What is he talking about? Is he in fact a male? Sarah wasn't so sure. Somehow it was hard to tell.

Sarah looked down at her own body. "This is not my body."

"You have a new, perfect one. My name is Javi," the person said. "I am your neighbor. Let me introduce you to your sleeping room. You have a soft bed, a closet, a bookcase of popular books from the library, a rocking chair, a prayer rug, a Buddha statue, and a standing Christian cross, which we know is common on Earth. You may have already had glimpses of memories from your home planet, but with time, more memories will refresh you."

They walked to the next room, a small kitchen with a table and chairs. The décor was sparse suggesting there was not much reason to spend time indoors. Javi escorted Sarah out the front door to view flowers covering the entire yard. The house was a cute little white-clapboard with a thin shingle roof. "This is my house?" asked Sarah.

"Yes, of course."

Sarah looked up and down the road to see a hard packed dirt path wind past small houses, giving the town an old English look. The yards were filled with small trees featuring bright green or blue leaves alongside flowers of various sizes. Friendly people walked along greeting each other, and said "Welcome" to Sarah.

"Bodies from Earth enjoy running," said Javi. "Let's run the mountain, Mt. Hope, so you can see your new home."

"Run the mountain?" Sarah questioned. "You mean *up*? Can we do that?"

"Oh, yes!"

Javi began running, and Sarah followed his lead. He raced faster and faster. Sarah was amazed she could keep up. He skipped over a tree. *Hmm*, Sarah decided she could trust this move, so she jumped up and crashed into the top of the tree. Pine needles scraped her skin.

Javi looked back. "Just keep jumping!"

They hopped up cliffs, bounding higher and higher until they reached the top of Mt. Hope. Sarah was awestruck. Resta was the most beautiful place she had ever seen. The mountains near and far were majestic, their peaks glistening with silver snow and their cliffs striped with sparkling waterfalls. The skies were bright blue but changed occasionally depending on the Brightness of the Beyond. Hints of purple and orange shimmered in the sky far away. The hills and valleys held a wealth of farmland and deep forests, animals on the ground and in the air, and farmhouses, mansions and small houses.

Miles away, by a small white building, two people were locked in an angry conversation, as indicated by their body posture and facial expression. Sarah glanced toward Javi who said, “Yes, you’ve noticed our acute vision. That is Pastor on the left and Voder on the right.”

“What are they arguing about?” asked Sarah.

“Pastor wants to continue managing the town, but Voder repudiates that plan,” said Javi.

Sarah noticed a dark wavy line encapsulate Voder’s body then morph into spikes. She found this event so unnerving, she looked away. But curiosity got the better of her, and she glanced back to see Pastor jump up on the roof, scream something while shaking an angry fist, and drop to her knees. She pointed down the road for Voder to leave. Voder bellowed something and departed.

Sarah shook her head and gave Javi a questioning look. Javi did not explain Voder’s behavior but swept an arm out over the stunning view—a valley where a silver river ran and herds of tigers and deer grazed. Such bright, bold colors everywhere. She closed her eyes tight then opened them. Was the beauty still there? Yes, it was. Her eyes teared up at the sight of a place in the universe as heavenly as this, and her heart filled with an exhilaration never felt before. Below her, spread throughout a large valley, was the town, Sedona, her new home in Heaven.

"You see," said Javi, "all of Planet Resta is as beautiful as this, both in physical beauty and spiritual beauty. Your family, friends, and all Earthlings are happy and loved beyond measure."

Javi's blue eyes and aura were warm with love and laughter. Sarah had heard of auras before, the entrancing glow around one's body, but never saw one, until now.

"Follow me!" yelled Javi.

They jumped off the mountain and soared like eagles down to the valley. They swerved right and left, boing-ed around the town as if on pogo sticks. Javi landed by his house and Sarah went *splat* in front of hers, misjudging her final leap.

She was giddy. "How are we able to fly? Do you have some special powers?"

"No," said Javi. "Every person is able in these perfect bodies." He explained the genders: no sexes, no man or woman, just "asexual person". No sexual activity. But, Javi explained, some people thought of themselves in their own minds as they used to be on their home planet — "he", "she", or "ke" (if something else). On Resta, the pronoun "ke" was used frequently because it meant neither male nor female, or that gender was irrelevant. The major possessive pronoun was "kes" but people also said "his", "her" or "their". Yet Sarah still thought of herself as a "her" even though she was flat chested and had no hips to speak of. *Ke, ke, kes, kes… this would take some getting used to.*

Sarah took this time to really look at people and was surprised to find they looked basically the same: smooth, dark tan skin, round face, high cheek bones, flat narrow nose, full lips, and large ears that could turn outward. Besides a brow ridge, there was a smaller ridge running from the eye to the ear. Everyone had dark hair that could be cut to whatever length was desired. To Sarah they all resembled males rather than females.

Sarah asked Javi, "Where are you from?"

"I'm from the planet Algenia in the Milky Way Galaxy, proximate to Earth, astronomically speaking. Life was extinguished on our planet by seismic activity."

"You're an alien! How many people came to Resta?" Sarah asked.

"All 50 million."

Sarah noticed that Javi's lip movements did not match the words she was hearing.

"What's your native language?" she asked.

"Algenian, of course. But you automatically translate my language into your English. I automatically translate English into Algenian. It's one of the senses of our bodies."

"How many senses do we have?" she asked.

"Many. Too many to learn about now. You will discover them soon enough, and I will help if you so desire," he said.

"Oh! I'm an alien, too. That doesn't seem right. A Caucasian, English-speaking woman being an alien. Are there many people from Earth here in Sedona?" Sarah asked.

"A couple. I enjoy talking with them. They are always eager to discuss family, employment, and recreational activities. I believe they lived only to retire and play golf. There is a golf site and many other activities around town if you wish to participate."

Sarah smiled. "Thank you. I'll have to take a look."

"I am impressed that many Earthlings strove to develop spiritual lives by attending church or temple. This is a laudable practice. On Algenia, every soul prayed at the temple three times a day, thanking the One Great One for all the blessings given to us. We, who deserved nothing, had everything required to be happy." Javi put an arm around Sarah and said, "I am glad you are present. I am fortunate to have a new friend."

Sarah felt such love in that simple hug, warmed that she had made a friend so easily. Sometimes, Sarah made ironic comments, and it would take a while for people to get used to it. But now, she would try to be kind and open.

They decided to go to the Tall Grass Restaurant for dinner, a small diner that specialized in celery, rye, and oat dishes flavored with herbs whose names Sarah couldn't pronounce. Javi selected Sautéed Rye Soup, and Sarah started to order the same.

"No, no!" said Javi. "You order a different item, and if you don't savor it, we can exchange."

Sarah selected Oat Cake with bell pepper and various herbs.

While eating their equally delicious meals, Sarah and Javi discussed food and daily activities on their home planets. Javi had been a teacher of agriculture and volunteered at a hospital. He did not mention any family. He liked living in Sedona. There was no government, no laws, no angry people and all the freedom and surprises a society could ask for. Coming from many, many planets that were destroyed, residents now had wonderful new lives.

Sarah asked why he said there were no angry people when they had just witnessed Pastor and Voder in a spine-chilling harangue. Javi explained that it was inconsequential. No one took their arguments seriously.

Sarah described Earth where she had been a woman, happily married, an elementary school teacher. But it was too short a life when the asteroid hit. As she recounted the story, more and more memories surfaced. She loved her job and put all of her energy into it – providing individualized attention to students who needed it, but also keeping her entire class on track. She made sure she smiled at each student and complimented him multiple times a day. She had huddled with her husband and other family members in their basement waiting for the end: a tsunami, hurricane force winds, starvation, or whatever. Even though she talked about her family, she didn't become teary-eyed over missing them and briefly wondered why that was so. "Do you know where my husband is?" she asked.

"Somewhere on Resta, but I don't know exactly where. He is completely happy and well-cared for," said Javi.

I am richly happy, too, she thought. *Is it the environment, the aroma of the food, or the people? It's so peaceful and calm here. Heavenly, hmm.*

In the candlelit restaurant, Javi's warm blue aura shifted to gold. Sarah wondered if he, or ke, back on Algenia, was a man or a woman.

CHAPTER 2

VODER AND ZAC

In the early morning, the grass and trees were particularly beautiful, each blade or thin leaf dotted with glistening dew. A herd of deer grazed in a park at the edge of town while greenish-colored finches flitted overhead. Two people sat on a park bench, relaxing, their legs outstretched and feet crossed.

“Query. Did we all die?” asked Zac, although it was more of a statement.

“Regret. Yes. I’m sure of it,” said Voder. “Before the war, Frond was beautiful. I wish you could’ve seen it. Purple skies. Vast savannahs, gently falling rains. Plenty of meat for all. The cities were clean and the economy stable.” He shook his head. “I should’ve been able to stop the war. To let it escalate to… well… it was inexcusable.”

“Pride. I think you were the greatest general the world ever saw,” said Zac.

“Humility. That is a generous statement,” said Voder, but thought *You don’t need to tell me--I know I was the greatest military leader in the history of the world.*

Zac elaborated, "Elation. You bombed those eastern barbarians. You established the world-wide land use pact. You obliterated the military threats to the pact. You suppressed the slave rebellion. Everyone admired you, such a sophisticated man tackling the world's dirtiest problems practically single-handed…" Zac's voice now crescendo-ed, "It made you the most honored general ever."

"Humility. Oh, God. I should've been able to stop that war…" said Voder, falsifying humility.

"Consoling. You had no way of knowing that the other nations reached nuclear capabilities. Don't blame yourself. There's nothing you alone could've done."

"Grief. But I do blame myself. The pain of watching my family die of radiation sickness was…"

"Sadness. I know. It was horrible," said Zac. "I can't imagine having to watch my wife and children die." He thought for a while. "Excited. You were clearly God's most important servant, and I think you still are. You have been given a second chance. Any planet, including Resta, with a diverse population has the potential for civil wars bringing destruction and massacres. We will continue to persuade the town to agree to a command center with you as the mayor or general. Or we could set up a selection or election process, whatever it's called, to make you the mayor. Then you will rest easy that peace will prevail. Your nightmares will end and your fears diminished. All the populations will be grateful. Sedona will become forever secure. If I may speak so boldly, you don't have to structure society exactly as it was on Frond. You can elevate the status of women to gain their support, even though we know there is no equality."

Resta will never suffer, thought Voder. *I will unite this planet. The military I amass will include military personnel from other planets. Males will predominate. It is only natural.* Voder took a breath and tried to convey a hint of a smile, although he could never really pull that off. "Happiness. Thank you, Zac. You are a dear friend. I wish I had known you on Frond."

"Appreciative. You had too many under your command to know a low-rank like me."

"Encouragement. I think you would've had a prominent military career," said Voder. He continued, "Frustration. But, here we are. Doing nothing of any importance. Wielding a hammer, sharpening a plow, playing a game. What's the point?"

"Encouragement. The point," said Zac, "is to discover the point."

Voder just rolled his eyes. The boy meant well.

They sat in silence while Voder's head tilted back as he remembered his final days with his family. He was so proud of his son, developing into a strong young man. He missed his wife, of course, her devotion and accolades. He never hugged or kissed her—that was not the Frondian way. But at the end, he held her as they hunkered down in his pretty little tin house with plenty of meat and purified water. But that was little comfort knowing that death by the blast or slower radiation sickness would be their fate. His wife prayed and recited verses from the prophets' writings, explaining, to the best of her ability, what they meant. His son was brave but angry. They heard the ferocious winds and huddled together. His wife and baby girl succumbed first. Then his son was lost. Then Voder himself.

"Query. Would you like to attend church with me?" asked Zac, inadvertently shaking Voder back to the present.

"Disgust. No. Yes, I suppose. I hope no freaks from Lanet are there."

"Encouragement. There may be one, that's all," said Zac.

"Indignation. I hate it when they get inside my head. They tunnel through me, trying to destroy my thoughts and my dignity. Those people are going to be segregated. I have no need of them. And the Narths, they stare at me like hungry monks hoping to see a new Frond emotion. Something they can pity in their own self-righteous way," Voder declared.

Zac didn't know what to say. They got up and headed down Main Street.

From amongst the herd of quiet grass-munching deer, Sarah's head slowly rose. She tried not to eavesdrop on these two men, but she couldn't help it. One was Voder, the general she saw arguing with Pastor. The other was Zac, his subordinate. *Suppression of a slave rebellion? Bombing barbarians? Calling people freaks? What kind of people were they?* she wondered. *And a command center? What was that? And they want to make Voder the general of the town? Are we going to war?*

Sarah watched them leave. The General stood tall and stiff, like he was in a starched uniform, his robe cinched tightly around his waist. Sarah did not see an aura around him, but there was that faint wavy line, maybe. Zac seemed a little looser, relaxed in his robe, and having a pleasing greenish aura around him which wafted toward the General. Sarah watched the General march and his friend shuffle down Main Street. They turned on a small lane and entered the little white building with stained glass windows on every face. It was not tall, the way churches on Earth were, but rather short with no steeple. She hadn't been to the church yet, but didn't want to follow them in. So she peeked in the doorway. A person Sarah recognized as Pastor, knelt by Voder and Zac and prayed. Then Pastor sat back in a chair and glared. *What is she looking at?* wondered Sarah. Some harsh words were spoken but she couldn't hear them, so she shrunk away from the church.

CHAPTER 3

ARON

Sarah looked through her closet to select a clean robe following her communing with the deer. She donned a black one, looked in the mirror and saw the Wicked Witch of the West. So she changed into a light green one. She took a good look at her face and confirmed that she did bear a striking resemblance to Javi. *Is this the image of God? Maybe. If I died on Earth, traits should have been eliminated. Now I should express Godly traits*, she surmised.

She sauntered into town to peer into the restaurants and shops. The streets were filled with friendly people who introduced themselves or touched her hand as they passed. She occasionally saw two people walk up to each other and touch hands, wriggling their fingers, bending their wrists, and tapping fingertips, all four hands busily communicating in a high speed sign language. Once she saw two people walk up to each other and pass through each other. Wow! But mostly, people communicated orally, and she heard it in English. There were lots of smiles and hugs. *These people are so kind*, she thought, believing she could easily grow to love this place and these people. She truly wanted to belong to this wonderful community since this is where God put her. It affected her in a surprising way—like wind was blowing through her mind, expanding it and opening it up. She thought she heard voices; no, whispers of words. How calming.

Main Street was lined with robe stores, wildlife supply shops, shampoo and lotion shops, game shops and bookstores. Sarah noticed that all the restaurants were vegan. There was a town hall, library, and theater. In a park she noticed little round, furry balls, rolling about. *Hmm, a new kind of animal*, she thought.

She passed by some people talking in one word sentences.

"Food. No," said one to another.

"Now. Food. Later. Play," said the other.

"Bull. Tiger. Farm."

"Yes."

I wonder where they are from, she laughed. *They sound so silly.* She walked down a lane to find a mill grinding grain into flour. The mill structure was roughly hewn boards making three walls and a large roof. Empty carts lined a large wooden platform. Next to a large pile of sacks was a giant gadget which, Sarah thought, filled the sacks with flour. She watched the millstone for a while, a huge wheel-shaped stone on an axis, but was unable to ascertain how it turned unassisted. There was a wooden pole emerging from the side, as if a donkey could be hooked up and encouraged to walk with the offer of food. But there was no donkey.

She was startled by a person who came up behind her and gave her a firm hug.

"Welcome to Mill Creek! My dear Soul, it is an honor to greet you on this beautiful day, beautiful day! The Brightness of the Beyond is great and your joy will be complete!"

That was by far the happiest greeting Sarah ever heard in her life. She smiled wide and turned, expecting to see a grinning old man, but of course it was a person whose face and body looked similar to her own. "Th-thank you! My name is Sarah."

"And I am Aron, the miller here. Unfortunate we have not met before, not before this. Are you new to Resta?"

"Yes, I am," said Sarah with a gentle breeze wafting through her mind. "I am curious what turns the stone."

"An underground river powers this mill to provide all the meal and flour for Sedona, for our little town! We are grinding wheat today. Would you like to place an order?"

"Oh, no," said Sarah. "I don't have plans to do any baking. I'm just learning my way around and meeting new people."

"Well, I am delighted, so delighted, you strolled down this lane to meet me. Would you like a job?" He laughed and gave her another great big hug, lifting her briefly off the ground.

"A job?" said Sarah, surprised at the question and the lift. She hadn't thought about finding a job. "I didn't realize I'd have to work, but I guess I should." It took a moment to formulate a response. "I don't think I'm ready to commit to a job. I'm still trying to learn my way around. But thanks for the offer!"

Aron was glad to meet the newcomer. He didn't have to delve behind her eyes too much to know that she was an honest, open person, no threat to anyone, and no preconceived notion how Planet Resta could be of benefit to her.

Sarah turned and walked quickly toward the town. *What a happy person! How can a person be so happy? Is he that happy every day?* She loved the way Aron spoke in sing-songy phrases, a rhythmical cadence so easy to listen to. And did he have an aura? Yes, she thought so.

A person jogged past her toward the mill. Thrusting arms into the air, he said to Aron, "I can't find Paul. Curses! I can't find Paul."

Aron gave instructions to the other mill workers, and several of them headed off in different directions. Sarah wanted to help, but she didn't know the person and didn't really know the town. So she headed to Main Street and entered a bakery. Ahh, the wonderful aroma of freshly baked bread. She asked for a chocolate croissant but shook her head when she realized she didn't have any money, then remembered from the restaurant the previous night that there was no money. Good.

Chapter 4

Mud

Sarah continued along Main Street toward a beautiful meadow where a dozen people were birdwatching. Oohing and ahhing, they pointed to various birds they spotted, admiring the coloration and songs. But more than that, they were excited to see the birds, as if they had never seen one before.

One of the people glanced up at Sarah standing there and quickly looked back toward the meadow.

Sarah said, "Hello. I am Sarah. You have so many beautiful birds here."

"I am Mud, the bird caretaker. We do not have the birds. They have us," explained Mud in a soft voice. "They flit about this meadow hoping we'll bring seeds. But they have all the seed they need without us. Resta, in all its abundance, provides for every animal."

The group gasped as they saw a very large, screaming bird tear across the field and land in a bare spot. Sarah expected to see the hawk-like creature pop up with a mouse in its beak. But instead, a chipmunk stuffed a large tuft of hay into the bird's mouth. The bird then flew away.

Sarah was impressed that this bird, resembling a hawk, didn't eat the rodent. Sarah chuckled at the surprise—a vegetarian raptor.

The rest of the group introduced themselves to Sarah, welcomed her to the meadow, and invited her to a talk on mountain birds at the bird supply store that evening. She said she would plan on attending. As she turned to walk away, Mud joined her and thanked her for stopping to talk.

"Where are you from?" Sarah asked.

"I'm from Earth!" she said. "Can't you tell?"

"You're from Earth, and your name is Mud," Sarah said with a grin.

"That's right," she said, without a smile.

Sarah knew that Mud did not speak English because lip movements did not match the English Sarah was hearing. Sarah paused, wondering if it was appropriate to ask the next question she was dying to ask. She said, "Sounds like a feminine name."

"Yes, it is, but we are not concerned about that now. You will discover that gender is totally irrelevant. Our bodies are all the same. Our love for people is derived from our desire to be kind, patient, and accepting of all."

"May I ask one question? If the chipmunk was not eaten by the bird, how are rodent populations controlled?"

"Animals are not food. Animals do not reproduce, and their bodies do not age."

"Interesting," said Sarah. "Thank you."

Mud smiled a meek smile and nodded her head. Sweet, meek Mud.

Sarah continued walking along Main Street toward the farmland. She had a new appreciation of animal life and wondered if it was possible to keep a pet. As care-free as a child, she relished the thought of living on this beautiful planet. But shouldn't she be wondering about her husband? How can she be happy without him?

Mud watched as Sarah walked on. How much trouble would Sarah, a person from Earth, be? Perhaps Sarah would focus on herself and her own spiritual development and leave other people's problems alone. *Yes, that's what I pray for,* thought Mud. *If she can't do that, then please, God Almighty, get rid of her.*

As Sarah reached a huge farm, she looked across the waving grain and sat down on the road, grateful to have time to enjoy the farm's beauty. There were a couple of farmhouses, a wooden barn, and farm equipment possibly made of wood. In fact, the road she was sitting on looked a lot like wood. *How is that possible?* she pondered. *It must rot over time. Or maybe it is repaired frequently.*

Sarah was startled when a sudden rustling noise sounded from the tall grain. Her eyes grew large as a huge cat-like face emerged and twitched its whiskers. Sarah remained as still as a statue. She was in shock and didn't know how to get up and run. Maybe that would make matters worse anyway. Calmly padding out toward Sarah, a long tiger-like body encircled her from its nose to the tip of its tail, and laid down without touching her. After a while Sarah realized she was holding her breath. So she slowly started breathing as quietly as she could. The tiger, or whatever it was, closed its eyes and faintly purred, giving Sarah the opportunity to examine it. It had thin, washed out stripes on its yellowish body. Its white and yellow fur was sleek, and on its underbelly, white skin could be seen beneath the fur. No aura. No claws. But Sarah knew the claws could be retracted. She wanted to look in its mouth and examine its teeth to give her an idea if she was in danger. She reached over the creature without touching it and snapped off a stalk of grain. She held it in front of its nose, and it opened its mouth. She moved her head down close enough to see the animal had only blunt teeth, no meat-shredding teeth. She gently put the stalk into its mouth only to have it spat out. Sarah chuckled.

She whispered, "So you didn't eat Paul?" She quickly regretted asking that question because if the animal answered, she would've truly freaked out.

After a while, the tiger twitched, woke up and wandered off into the field. She watched it, wondering if it would turn its head around to look at her. Perhaps a wink of acknowledgement that it knew Sarah wanted to be friends. "That's really asking too much," she said out loud.

Sarah began wondering what her job should be. Tiger caretaker? She knew nothing about it. Croissant baker? That might be fun. A miller? Definitely not. She really liked meeting and talking with people. The town greeter? Yes. Perfect.

Later that day, Sarah went to the bird supply shop to hear the talk on mountain birds. It was fascinating! She learned about the uniqueness of each bird, its behavior, habitat and diet. And after her encounter with the tiger, she knew that every animal on Resta was precious.

Before the talk was over, a person walked into the shop and stood in the back. Mud, with her meek smile, walked over to the person and whispered, "We have almost finished here. Perhaps you could attend a future talk?"

"Yes, of course," said the person. Ke turned and left.

How odd, thought Sarah. *There was no reason that person couldn't have stayed for the last few minutes.*

At the conclusion of the talk, all were ushered outside to run the mountain. Sarah's 'run the mountain' skills were much improved, and she was able to land accurately at their destination – a cliff just above the tree line where rare small, swift-like birds lived on vertical faces. The birdwatchers, perched on a ledge well above the birds, admired their cigar-shaped bodies and how fast they could beat their stiff, narrow wings. Sarah was amazed that the birds could live on vertical faces. Just a one-inch shelf or a tiny plant poking out from the rock was enough area on which to build a nest. Similar birds had lived on Earth but not most other planets. Sarah once again appreciated how good her eyesight was when she could still see one of the birds fly many miles away. They observed the birds for a long time, delighted to learn about these awesome little creatures.

The people boing-ed down to the town. Sarah walked next to Mud and said, “Thank you for inviting me to the bird talk. I really enjoyed it. I’m just wondering why the person who entered late could not stay for the end.”

Mud whispered to her, “That person was from Frond and did not realize that the talk was about birds. Those people do not appreciate our animals.”

“But wouldn’t a talk at the bird supply shop be about birds?” Sarah said. “Were there other people from Frond at the talk?”

“No. I don’t encourage them. They really do not enjoy it. There are so many other fascinating activities for them in Sedona. And I do want them to be happy.”

Sarah thought about the strangeness of Mud’s attitude. Mud wanted them to be happy, but felt they needed help deciding how to live their lives.

Sarah took a deep breath, but then relaxed. She was not Mud’s mother and couldn’t tell her what to do. She just said, “Shouldn’t they be allowed to make their own decisions?”

“Shh,” said Mud. “Keep your voice down. It does not help to complain about them.” Some people in the group heard these comments.

“I’m sorry, Mud,” said Sarah, trying to lower her voice. “I’m not complaining. I’m just trying to understand. I would want everyone who didn’t know about birds to attend the talks.”

“You will understand soon enough. Those Frondians are controlling and manipulative. They may appear kind, but they are actually politically motivated and desire the subjugation of others. When you meet them, you will see. There is no room or reason for politics on Resta. Resta is the most blessed of all the worlds and must remain unblemished.”

Mud had escalated this simple conversation about a bird talk up to a whole new level. Sarah took a moment to absorb Mud’s comments and then apologized for starting such a contentious conversation. Mud said it was understandable, given what Sarah had witnessed, and told her not to worry.

Sarah strolled about, thinking of the reality of living here without her husband. If she didn’t think about him, then yes, it would be okay. Maybe, if she could keep busy enough. She looked deep in her heart, and she sensed that he was completely happy.

She did not see Pastor watching her from the roof of the little white church.

CHAPTER 5

THE CONCERT

The next morning, Sarah walked by a restaurant and was drawn into it by the tantalizing aromas of fruits and biscuits. While eating a lemon crepe, she noticed that the table was too polished and shiny to be wood. It had a pleasing opalescent quality, almost like the nacre of the inside of an oyster shell. She tapped it with her fingernail and heard a slightly metallic sound. There was nothing like it on Earth.

"It's from Algenia," a voice said behind her.

"What?" Sarah turned to see her neighbor Javi.

Javi explained that the table was wood coated with an aluminum-like mineral similar to what was found on Algenia. The coating provided everlasting resilience making it pleasing to the eye. The utensils were forged from another element found on Algenia. Sarah began realizing that Resta was composed of the best of the best. Javi concurred.

"Do you have a planned activity this morning?" Javi asked.

"No," said Sarah.

"There is a concert in the theatre which you may enjoy. Would you like to attend?"

"Yes, I would. Very much. When is it?"

"It will commence in a moment."

Javi laughed as Sarah inhaled the crepe, and they dashed down the street to the theatre, entering along with a large crowd. The theatre was much larger than its outer appearance indicated. There were balconies, chandeliers, carpeted aisles, a large opalescent stage, and over a thousand seats, Sarah guessed. The walls were panels of various varieties of wood, highly polished. She and Javi moved to the center of the theatre and sat in rich leather seats, which Sarah admired and thought they may not be leather at all, but something even better. On stage, there were about 50 chairs with a small table in front of each one. On each table was an instrument, each with a different shape and size. Sarah couldn't identify a single instrument. One looked like a trumpet, but instead of one horn it had four. One looked like a drum with pipes running through it. One looked like a tin can with gold tubing wrapped around it. She saw nothing that looked like a violin.

Sarah was so excited as the musicians filed in. When they arrived beside their seats, they knelt on one knee and waited. Then the conductor walked in, dressed in a brilliant blue robe, like royalty, and a gold tie around the waist. Ke stepped up on a low platform and faced the orchestra. Ke did not face the audience, and there was no applause. The musicians stood and picked up their instruments. The conductor, with raised arms and no baton, moved kes hands to start the musicians playing. Some musicians blew into their instruments with their mouths. Others waved them in the air. And others tapped on them. Sarah could make no sense of it.

But the music! It was exquisite! Sarah's mouth fell open. Javi looked at her and started laughing. Realizing how ridiculous she must've looked, she put her hand over her mouth to keep it closed. She shut her eyes, letting the music fill her heart and mind. How could an orchestra play more beautifully than those on Earth? How could a work be more beautiful than Beethoven's Ninth Symphony? And yet here it was. Her heart felt lifted as the melody evolved and the instruments' tones blended. She opened her eyes to try to follow which instrument was making which tone, but it was pointless. There was no analyzing this moment.

Sarah was spellbound throughout the entire three hours. The conductor used no score, and kept arms raised the whole time. After the concert ended, with no applause, the conductor and orchestra filed off the stage. The audience then cheered and left. Sarah and Javi stayed seated.

"I don't want it to end," she said. "That was the most beautiful music I've ever heard."

"Happiness is coming out of your ears," said Javi, smiling.

"What?"

“That is an expression on my planet, but it applies to you.”

“I would say ‘I’m so happy I could cry’,” said Sarah.

“Why would you cry if you are happy?”

“Tears of joy. It’s a mostly female emotional thing. Is this orchestra from Algenia?”

“Oh no,” said Javi. “We had no musical instruments.”

“No musical instruments? Did you have any music at all?”

“We sang. We sang praises to the One Great One. I still do, with my voice but mostly in my heart,” explained Javi. “It is very satisfying.”

“Query. Would you like to hear it again?” someone asked.

Sarah and Javi looked at the stage to see the conductor with a big grin. It was Voder, the General!

Sarah was speechless for a moment then said, “That was the most beautiful music I’ve ever heard! Thank you so much.”

“Cheerful. This has been my pleasure! And welcome to Resta. My name is Voder. This is quite a delightful friend you have made, Javi. Thank you for bringing her here.”

“Voder, this is Sarah, my wonderful neighbor,” said Javi.

“Wh-what is the name of this piece of music?” asked Sarah. “Did you write it?”

“Pride. Oh, no,” said Voder. “A composer from my home planet Frond wrote it. It is called ‘Spirit of Love’.”

“I must confess I could not identify a single instrument nor the sound of any instrument.”

“Encouraging. Nor should you. But, if you love music, you will gradually learn about them. Who knows? Perhaps you will be my next grugru player! Males play the best, but supposedly a female could be adequate.”

Then Voder said goodbye and departed.

Sarah furrowed her eyebrows. “Well, *that* was weird,” she said, with the emphasis on *that*. “Something very strange happened to me yesterday, Javi. I overheard a conversation between Voder and Zac and learned of their lives and the destruction of Frond. Voder was a great general, but to me he sounds like…” and then she whispered, “an evil dictator. And from that ‘a female could be adequate’ statement, he is also a chauvinist.”

“He is usually kind,” said Javi, “but forgets his manners upon meeting a new person.”

“No, I don’t think he forgets his manners,” said Sarah, clearly annoyed that Javi could believe that. “I think he needs to establish his dominance from the beginning, making me stay in my place.”

"I never knew that about males," said Javi. "Very interesting, but unfortunate."

"It's not interesting," said Sarah. "It's a sad comment on male dominance. It commonly occurred in various ethnic groups on Earth, and now I see it was present on Frond. Someone told me that people from Frond are manipulative and controlling. And now I've met a person whose soul expresses such beautiful music and yet wants to be in control militarily."

"One thing you must remember: our souls come to Resta along with our characters and experiences. People have flaws and regrets and even sorrows. With effort we can heal our flaws and bury our sorrows. But it does take time, so we still express our personalities and emotions. We come from different planets where different practices were acceptable in society, and where different character traits were considered more admirable than others. On Resta, we all live together and learn to exhibit the gifts the One Great One wants us to express.

"Well, Voder's militarism accompanied him here," said Sarah. "Voder and Zac want Voder to be the mayor or general of the town to ensure peace. They believe in war and population segregation."

Javi stared at her.

"What," she said flatly. She could tell Javi was stunned.

He took a deep breath and, with difficulty, said, "Voder may have been a dictator, but on Resta, he is a good person. He must not become a general. That is not a job."

"I don't understand why he thinks the town is in trouble," said Sarah. "Why would he want to be a military leader? That's what 'general' means. Am I understanding this correctly?"

"Yes," said Javi, sadly. "But I wish not."

Later that day, Sarah decided to run the mountain. She sat on a cliff perch, closed her eyes and hummed. She thanked the Lord for her new life. She prayed she could be as good a person as Javi, always pleasant, always helpful. When she opened her eyes, she spied Voder and someone she didn't know with a blue aura sitting on the ground far below her at the base of the mountain. She saw a wisp of blue aura extend out and reach for Voder. Voder's own dark aura shrank to nothing, and he stood up and left.

Sarah watched a person enlightened with a yellow aura sit down on the ground facing the blue aura person. Once again a wisp of blue aura extended out from its owner. But this time, the yellow aura also reached out, and the two wisps swirled together like a dust devil, an intimate entanglement of...*what* Sarah wondered. Yet another way that two individuals can communicate? Some unique form of telepathy?

Sarah could not see the individual subatomic waves, but they were there, mixing, blue plus yellow. She knew something was happening:

particles from each person were emitted and traveled along the waves, creating an energy, a spirit, which linked the individuals together. Sarah saw a bright blur between the people, and it was beautiful—souls without the body. A blur of joy. The dust devil lasted for quite a while. Sarah thought maybe this interaction could be a way for people to really care for each other, to love each other and to intimately help one of them with a problem without necessarily talking.

Then the yellow aura person got up and left, and another person, with a red aura this time, sat down facing the blue aura person.

But the red aura person, disappeared. Poof! Vanished. *What happened*? Sarah wondered.

“I’m over here,” a growly voice said.

Sarah, startled, looked all around. “Where?”

“Top of the tree to your right.”

“What are you doing?” she asked.

“Wondering why you spy on everyone,” ke said.

“Please don’t be angry. I’m not spying. I just find myself in situations where I can observe people interacting. Eventually, I’d like to meet and get to know everyone.”

“Don’t bother about me,” ke said. “I’ve already met everyone I care to.”

Then poof! Ke was gone again, this time reappearing seated next to her on the ledge. Sarah just about jumped out of her skin.

“Would you stop doing that?” she shrieked.

“I want you to know I’m watching you,” ke said.

“You can watch me, all you want. Can I ask who that person is …” she started but didn’t complete the question because ke poofed again. She continued, louder, “… person is with the blue aura? Is that your psychiatrist?”

I guess people need help everywhere, Sarah thought. Below her, office hours were over—the blue aura person was gone.

CHAPTER 6

JAVI

The next morning, Sarah woke imagining she heard music—Christmas songs that she and her friends liked to sing. *Oh I miss them*, she cried. With tears in her eyes, she dressed, ran next door to Javi's house and knocked on the door. He opened the door, saw her tears and said, "Oh, you are happy about something this morning!"

"No. I miss my husband and friends," she said as she threw her arms around Javi's neck. "Can you come over for tea?"

Javi sat at her kitchen table while she made tea and served gooseberry muffins she had picked up at the bakery the night before.

"I'm sorry you are sad. I did not realize this would be problematic for you. I think you will feel better with time because eventually your soul will touch that of your husband. Then you will feel this is your home. If there's anything I can do…"

"I don't know. I'm sure I will feel better if I can find his soul. Does that happen?

"Yes it does."

"But the real reason I asked you over is that I have so many questions to ask you, but I feel that I'm intruding if I do. Is my curiosity a positive or negative trait? I mean, how much can you tolerate from me?"

"I believe your curiosity to be mostly positive. I don't need to 'tolerate' you because I already enjoy your company. Ask away."

"Were you a man or a woman?"

"Neither."

"Neither?!"

"That's the correct translation. Your sense of translation is good."

"So that makes you a 'ke', not a 'he'. How did your population reproduce?" she asked.

"When a person died, the body automatically formed an embryo in the womb. The embryo grew using nutrients from the deceased body. After a couple of years, the baby was born and raised in an orphanage."

"So there were no mothers or fathers or even parents? Just orphanages? That's so sad!"

"We were all parents. Children received the best loving care and education. When they matured they selected their jobs and were trained well," Javi explained.

"Where did you teach? Was it agriculture?" Sarah asked.

"Yes, at the university," said Javi. "It was a good, stable job. In an agrarian-based society, I was respected and loved. But I was always envious of the miners."

"Why?"

"Being a miner was the most respected occupation. They overworked themselves to exhaustion and ill health. They were greatly admired for their dedication and sacrifice. However, I could not pursue that career due to my infirmities."

Sarah raised her eyebrows with a questioning look.

"I had deformed legs and undeveloped echo ridges, so I was unable to navigate the mines."

"What are echo ridges?" asked Sarah.

"These ridges that run from the ear to the eye," said Javi, running a finger along one side of her head. "They are the organs that enable the sense of echolocation. People on Earth didn't have this sense, but many of your animals did."

"And your legs? How did you move around?"

"I had a motorized cart which was adequate; but, I was still envious that others could walk, envious to the point of anger. Our society admired envy and jealousy, so it was good I felt that way. I asked the One Great One to redress my affliction and let me walk. He did not grant my request, until now. Since arriving on Resta, I have learned that envy is a sin. Daily, I pray to be forgiven."

"Redress my affliction?" questioned Sarah.

"Heal my failings," said Javi.

"Do you find you fit in, here in Sedona?" asked Sarah.

"Yes," ke said.

"How long did it take you to feel that way?"

"Immediately upon arriving. I could sense the spiritual nature of these people. I want you to feel comfortable, too."

"Oh, I do," said Sarah. "And I've started wondering how to spend my time. If I'm ready to find a job. I want to build a life here, with good friends and meaningful work. Other people have devised strategies for dealing with people from Frond. I suppose I can do the same. But let's go back to Algenia. You said you volunteered at a hospital?"

"Yes, I sat with patients who were injured in the mines. I read books or newspapers to them and talked about what was going on in the world. When they were too sick to go to the prayer room, I'd bring the prayer cloths to them and pray with them. If a patient died, I'd sit with the body while the fetus grew and sing to it."

"How many people died in the mines?"

"Many. The mines were very dangerous. There were many collapses. The day the mountain range collapsed…"

"What?!" interrupted Sarah. "The whole mountain range collapsed?"

"Yes, from pole to pole. Following a massive earthquake, it slid into the ocean and caused the violent tidal wave that traveled around the planet and covered all the lands. It traveled across the entire major continent. I was at the hospital when we were notified it was coming. I held my patient in my arms as the water crashed through the windows. When I woke up, I was here on Resta."

Sarah was stunned for a moment.

"Do not be astonished," said Javi. "The One Great One loves us even if we have just a tiny bit of faith. While being here, I have grown a great deal in my spiritual life."

"Did you look for your patient here?"

"No. Resta is large. There are trillions and trillions of people."

Sarah and Javi sat in silence for a while looking out the window at a gentle rain that had started falling.

"I am very sorry," said Sarah.

"Do not be sorry. We were always in the One Great One's hands. I am at peace and I have been blessed with a perfect body."

After a few moments of contemplation, Javi said, “Your steeped drink is very tasty.” Ke examined the muffin. “What is a gooseberry?”

“I have no idea.”

CHAPTER 7

ARON

Aron, the sing-songy miller, was pleased with the flour produced that day. The workers did a good job collecting the flour, sieving it, and bagging it. It was evenly colored, just off white, and silky smooth. Aron looked at all of the workers, slapped one of them on the back and said, "You're the best! The very best of all the creatures! Ha ha! This will make the finest cake in all of Sedona. Did you enjoy your work today?"

"Yes," said the worker, Dora. "I found it so satisfying, I hug myself! I know everyone will appreciate our efforts when they taste tomorrow's cake, tomorrow's delicious cake!"

"Would you like to deliver it to the bakeries?" asked Aron.

"Yes, I will make the deliveries." Dora whistled a pretty tune.

"Thank you, Dora, my dear friend, my dear friend."

Aron reminded everyone to pray for Paul's happiness. The workers cleaned up and went home.

As the Brightness of the Beyond dimmed, Conductor/General Voder walked down the lane toward the mill. Aron saw him coming and waited patiently, but expected not to enjoy the conversation.

"Query. My dear Aron," said Voder, "have you thought anymore about my proposition?"

"Yes, and I think it's a distressing idea. I'm not interested, not caring at all."

"Indignant. It is a very good idea and will eliminate all sorts of disagreements in our town." Voder concentrated as hard as he could to build a wall of steel around his thoughts in an attempt to prevent Aron from reading his mind. He hated that more than anything. More than stupid, simple people. More than bland, pointless past-times.

"There are no disagreements," said Aron, "other than me disagreeing with you. You have no right to impose your will over others, not even one. That is not kind nor compassionate."

Voder's eyes grew dark. His anger increased until his heart hurt: the pain unbearable, much worse than anything experienced on Frond. His pride caused his wavy line aura to grow denser and darker, like black storm clouds forming. Aron stepped back, watching Voder's changing appearance. Then, Voder turned and walked back toward town.

Aron felt pity for Voder whose self-righteousness was disgusting and sinful. Aron did not know how to stop it. There would be no peace if Voder's controlling ways were allowed to permeate the town. He would interfere in everybody's business, worm his way under everyone's skin to create an atmosphere of meanness and political tactics. Aron himself was already finding himself burdened by keeping Voder away from the mill and surrounding farms.

A person on Resta must know to repress inappropriate behaviors and allow the beautiful traits to prevail, as did people from the planet Lanet.

I wonder if Voder could have a stroke and die. What a terrible thought! Aron decided to try to be friendlier toward Voder in hopes of changing Voder's ways. *If only every person behaved like the beautiful persons of Lanet!* "Oh, Lord," prayed Aron. "Please calm Voder's heart to feel love and peace."

Aron looked up to see Sarah coming. They hugged as Aron said, "It is wonderful to see you. I hope you have had a joyous day, a splendid day!"

"Yes," she replied. "I'm finding every day to be joyous. Today, I am improving my vocabulary."

"Ah. A benefit of conversing with Javi," said Aron.

"I'm wondering how you are. I just passed that pugnacious Voder whose appearance was not normal."

"We had a disagreement," explained Aron. "It happens, just as night is sure to fall, and Voder is sure to breathe."

"Would you like to join me for dinner?" asked Sarah.

Aron nodded, and they turned into the first restaurant they came to – a large buffet with a hundred different fruits and vegetables in a hundred bright colors.

"I think this will become my favorite place," admitted Sarah.

They went down the line with their trays, Aron explaining some of dishes. "Try that one. It is an arula tart with a whole mume grain wrapper, my personal favorite. Try that one, too. It's a slice of apple pie with cinnamon and binter spice. Delicious, simply delicious." They sat at a table for two, and Sarah admired his aura—a bright green aura of joy.

They discussed flour for a while. Sarah asked Aron if they found Paul; but no, he was still missing. Sarah asked if other people had disappeared. Aron said that yes, he knew of two people from a planet who wanted to find other people from that planet. They could not find anyone here in Sedona so kept on walking. Then, there was one other person who wanted a home here, but Pastor said no.

Sarah wanted Aron to tell her all about his home planet. Aron was more than happy to tell her about Lanet, a peaceful, happy world with many different cultures and ethnic groups. People were either male or female who mated for life. They raised their children, worked at their jobs, participated in many athletic games, and cared for the infirmed. Churches were everywhere, at least one per street. Preachers spoke frequently, as did teachers. The rest of the people looked into each other's eyes and shared the silent commune of minds.

"You are mind readers!" Sarah concluded. "Oh." She remembered what Voder said—that people from Lanet were freaks, and he didn't want them in his head.

"I suppose you could call it mind reading," said Aron. "I don't think there is a precise Earth word for it. We don't practice it with everyone on Resta – many do not approve. For us, it is the most loving form of communication, most loving. To see into a person's eyes and mind is to truly understand somebody's character, desires, needs, and suffering. It is a blessing that enabled us to build relationships—strong, spiritual relationships."

Sarah thought for a while then said, "There must be no secrets. To be so exposed sounds so shameful. On Earth, we confessed our sins to God which was like not confessing them."

"So confession was thought to be ineffective?"

"I guess not. I never thought about it. At most we'd talk about them with our best friend. We were mostly too embarrassed to tell others the bad things we did."

Aron said, "True, we had no secrets, and we did confess our sins to the Lord, to our merciful Lord. But the relationships between people were intimate and sincere. There was no sarcasm, prejudice, or animosity. Instead, kindness, empathy, love, and humor dominated our societies, all of them. People from around the world traveled and mingled like family, a huge yet close-knit family."

"Oh, family! May I ask if you had a family?"

"Yes," said Aron with a glint of light in his eye. "I had a wife, 11 children, two pets and a top-of-the-line bubble. I bet your next question is 'What was my job?' I was a bubble manufacturer. Bubble, bubble, bubbles!"

Sarah laughed out loud. "What?"

"A bubble manufacturer. Our mode of transportation was by bubbles, bubbles, soft round vehicles controlled by our minds. They contained a seat or two, electronics, and safety features. No person was ever injured in a bubble, not nary a one. If two bubbles were passing by, the occupants could extend their hands out of their bubbles and touch palms. Then they could communicate, building their relationship. And relationship building was the highest goal on Lanet. Bubbles were a great achievement and helped us reach that goal, our glorious goal ordained by God."

"Oh, Aron," said Sarah. "Lanet sounds so strange and wonderful. But if so wonderful, why was it destroyed? Did God let that happen? What destroyed it?"

"It became uninhabitable, destroyed by de-vegetation when our growing population progressed to its expected outcome. Food production terminated, rather suddenly. Technology was very advanced, and we thought we were controlling every aspect of our planet to make it habitable for our increasing population. But we miscalculated...

everything! We were a prideful civilization, but did not know it until it was too late, far too late. With the plants gone, the atmosphere filled with carbon dioxide and every person and animal on land slowly fell into sleep. I don't know what happened to life in the ocean. I'd imagine it died off, too, leaving all of Lanet… dead." Then he chanted, "Dead, d-e-a-d."

They spent the rest of the meal talking about life on Resta. "I love all this food," said Aron. "On Lanet, at the end, we ate one food pill a day. Here we can eat all day long! There is so much variety. And it is amazing to spend so much time walking. It feels so … natural! I don't miss my bubble at all!" He actually rose from his seat and walked around the table to give Sarah a hug.

Sarah began to feel the breeze in her mind, lifting her spirits even higher. "What I love," she said, "is how we all look alike, down to our dark hair, height, and weight, yet I can distinguish you from Javi and from Mud. How is that possible?"

"You have noticed the small differences between us: the spring in our step or how fast someone talks. Maybe you even see if someone raises an eyebrow or curls the lips. But mostly it is because you look into people's eyes to see their minds and souls, their unique minds and souls. You are very perceptive and want to see their auras and spiritual nature, and so you do. If you happen to see someone from Earth who you loved, you'd instantly recognize that person."

"On Earth, we always said 'heart and soul', not 'mind and soul', that what a person feels is in his heart." Sarah wondered if he was reading her mind.

"The heart is just a noisy muscle. Thump-da-thump. On Lanet, we believe, I mean we *believed*, that the eyes were the keepers of feelings and the pathway to the mind and soul."

"Does a soul have feelings other than love?" Sarah asked.

"Yes. But the greatest is love. It encompasses every compassionate, kind, joyous, gentle inkling we possess. It is deeper than anything we have ever known while in our mortal bodies."

"Can a soul live outside a body?" asked Sarah.

"Yes, as an aura or a stretch – streeeeetch of an aura," replied Aron.

"Can it live completely outside a body?"

"I believe so," said Aron, "but I've never felt it, seen it, sensed it."

"Last question," said Sarah. "How do souls express love and feel happiness?"

"They *are* love, they *are* happiness."

Chapter 8

Church

Sing-songy Aron loved his dinner with Sarah. He felt her pure heart, truthfulness, and open, loving thoughts, full of caring and joy. She didn't have to have peace in her to be satisfied, but rather preferred stark honesty. She was just what this town needed. He invited her to church for an evening worship service. She was delighted to go.

About 100 chairs filled the church and faced a marble altar covered with an embroidered white runner. Behind the altar hung a large wooden (or perhaps not wooden) crucifix. Aron and Sarah sat in the front row. There were about 10 individuals there. Pastor came out and stood beside the altar. She was dressed in shiny white robes and had a pale purplish aura around the head and hands. Aron saw Sarah smile and knew she immediately felt the love in Pastor's heart.

Pastor said, "Let us pray."

Sarah hung on every heartfelt word she said, and it filled her soul, just as a sail is filled with wind. Sarah looked at the altar glowing brightly, a whitish color with streaks of gold through it. Pastor said the phrase 'the glory of the Lord', and Sarah understood for the first time what that meant. Next, Pastor said the phrase 'Praise the Lord all you who are worried', and the congregants repeated it. Then she said another phrase, and the congregants repeated it. This continued for about a half hour. Then everyone began praying their own prayers to God, either mumbling or quite out loud, all at the same time. It seemed rather chaotic to Sarah, but she managed to focus inward on her own thoughts. At that point, Sarah realized she did not have a strong faith. Her little faith allowed her to believe in Jesus and Heaven, but that was all. The service concluded with Aron and three additional congregants singing a beautiful hymn. At that time she committed to think about becoming a three-dimensional spiritual person.

Aron introduced Sarah to all the people in the little church. There were lots of hugs. She asked how often services were held. Pastor said that whenever two or more people walked into the church together, a worship service was held. If only a single person entered, it was a time for meditation and prayer. If many people wanted to meditate, there was a room downstairs."

Sarah then asked what was being done to find Paul. No one responded. Not like it was someone else's responsibility, but more like it was a ridiculous question. Sarah repeated the question with her eyes, but still no response.

Out of the blue, Aron said, "I have a proposal. I nominate Sarah to be mayor of Sedona."

Sarah's jaw dropped as everyone nodded, "Yes, yes, good idea."

"I'll post an announcement on the town hall door," said Aron.

"Yes, yes, good idea," they said again.

Aron looked at Sarah. "Clearly, you are the obvious choice."

Pastor saw a fearful look develop in Sarah's eyes and told her that she, Pastor, had been told that Sarah was to be an important part in managing the town.

"But Voder wants to be the mayor," said Sarah.

"Voder doesn't even have a soul," said Pastor. "Well, sorry, that's not totally true. He has a pea-sized soul."

Is that the kind of thing a pastor should say? thought Sarah.

"Who told you I was going to be managing Sedona?" asked Sarah.

"God told me, of course," said Pastor. *"Multiple times."*

Aron and Sarah walked toward Main Street. Aron said, "Sarah, I have a confession to make. I lied to you."

"What?"

"A lie of omission. I neglected to tell you that I was a criminal."

Sarah was stunned, again. "You were a criminal? I don't believe it."

"I assure you, I do not lie. Well, not much. I am sorry to startle you. Do you want to hear about it?"

"Do I want to hear you shatter the image of a man I just learned to admire? Sure," said Sarah.

"My crime was industrial espionage. I stole the plans for a new generation of climate controlled housing units from a competitor."

"How could you do that?"

"Oh, it was fun! I had a great time, a delightful time! It wasn't really sinful."

Sarah stood, dumbfounded, looking at Aron in disbelief.

"It's okay," said Aron. "Please don't look at me like that. I asked Jesus to forgive me, and he did. I take responsibility for my actions, and, anyway, some crimes are not as bad as others. I'm asking now that you forgive me, too. Please forgive me."

"For stealing or for not telling me the whole truth?"

"For not telling you the whole truth," said Aron.

"Yes, of course I do. All that stuff about relationship building, was that a lie?"

"No, that's the truth, the absolute truth. People on Lanet knew I had an impure heart, but they had a difficult time detecting what was wrong with me. I only had to maintain the charade for a couple of days. Immediately after the theft, I was caught."

Aron saw a tired, 'I've-had-enough-for-one-day' look on Sarah's face. She held up her hand and Aron placed his palm on her palm. She noticed his aura was gone. He smiled, and they went their separate ways.

CHAPTER 9

OLD SOULS

On the back porch of Sarah's house sat a large porcelain bathtub about sixteen feet square with three steps on the outside going up and three steps on the inside going down. She turned on the water, squirted in a little bubble bath she had picked up at the lotion shop, disrobed and stretched out in the warm water. Heavenly.

"You do realize that's a baptismal pool," said Javi from the adjacent backyard.

She burst out laughing. "I had no idea. Are you pulling my leg? You are, aren't you?"

"Pulling your leg?" said Javi, chuckling.

"That means 'are you teasing me'."

"No, I'm not teasing you. English incorporates the funniest expressions," Javi said.

"Is it a sin if I stay in for just a little while longer?"

"No, it's not a sin, nor a crime, nor an ethics violation," ke said, laughing again. "It's just funny. I'll wait for you if you want to go eat." And ke went inside the house.

Later, at the large fruit and vegetable buffet, Sarah and Javi sat down for a lengthy breakfast.

"You are an agriculture teacher. Right?" asked Sarah.

"Yes, just as on Algenia."

"Is this the work God prepared for you?"

"I hope so," said Javi, laughing. "Otherwise, I'm wasting my time."

"Is there a school?"

"No," Javi replied. "I visit the farms when requested to teach one-on-one, diagnose problems, and help find solutions.

"I guess you heard that Aron nominated me to be mayor, and God told Pastor I would play an important role in managing Sedona. And she said that Voder doesn't have a soul, just a pea-sized soul. How can she say that? Is she really a pastor?"

"Yes, I believe so. How do you feel about becoming mayor?"

"I don't know. I'm honored but nervous. I don't want to spend the rest of my life fighting with Voder and Zac. I still feel new here, and I certainly don't know all the issues and what a mayor would be responsible for. I wish I could visit another town and talk to their mayor."

"That sounds worth pursuing," said Javi. "Your desire to become affable with everyone is a commendable quality."

"I don't feel affable with everyone," said Sarah. "I was critical and bossy with Mud. I made fun of someone I hadn't been introduced to, and I don't like Voder and Zac. I'm not off to a good start."

"I still believe you'd make a fine mayor. I can't imagine you taking bribes."

Sarah chuckled. "Why did you say that?"

"On Algenia," said Javi, "the government was very corrupt."

"Well, it's a good thing there's no money here," she said.

Sarah ate some strawberry-like fruit with a rich flavor. "Hmm," she said with her eyes closed. "I agree with Aron: the food here is wonderful. Maybe I should be a cook."

After a moment, she asked, "Where is Pastor from?"

"I don't know," said Javi.

"I suppose Paul has not been found," said Sarah. "Did you know him?"

"Yes. Ke loved working at the mill and at the winery. We played square ball frequently. I thought ke was happy, so I'm surprised ke would leave Sedona."

"Where would ke go?"

"South along that road, or north to I-don't-know-where, or over Mt. Hope perhaps, or across the hills and farmlands to those mountains, or down the river to I-don't-know-where," said Javi.

"May I ask you a very personal question?" said Sarah.

"A personal question about geography? Yes," answered Javi.

Sarah grinned at the little joke. "Is an aura the soul?"

"Not quite. It is a bit of the soul that readily makes itself available to love. A bit of spirit or energy."

"Are you able to expose your soul to God when you pray or meditate?" she asked.

"Hmm, *that* is personal. If I answer truthfully, it may appear that I am bragging. Well, you already know that I have a problem with envy. Since coming to Resta, I have learned that if I ignore my ego, then my envy becomes desire. I can desire certain behaviors or activities. That is good."

"I've never thought about that before," said Sarah, blown away by how spiritual ke was.

"And when I completely ignore my ego, totally, then my soul is free to love. That is what God desires—to see our loving souls. So, yes, I have learned to meditate and empty myself so that my soul is exposed."

“Javi, you are amazing and terribly confusing! You are the holiest person I have ever met. I should’ve gone to church more. I’m going to have to think about this a long time.” She put her head in her hands.

“You ask the right questions,” said Javi. “When your soul becomes older, that is, you spend more time recognizing it, you will find it is the way God wants it to be. Your soul will be free to join the Unity of Souls, a spiritual place to love others. You will find your husband’s soul there, possibly soon. You will discover that souls touch each other through a mingling of essence. Souls see each other, but there are no eyes, just an intimate perception. It is the way that people of different religious backgrounds are able to love each other. Souls express their love for God and each other, but there are no mouths, no voice, rather the joyous exaltation of spirit.”

“Is the Unity of Souls where God talks to people?”

“Yes,” affirmed Javi.

“Thank you for helping me. I don’t know what I’d do if I didn’t have you.” She looked deeply into kes warm eyes.

“Sarah,” ke said, “I have enjoyed your friendship very much, but I am concerned that you are treating me as though I were male.”

Sarah immediately said, “I don’t mean to. I actually feel that everyone looks like a male. I’m sorry. I’m just trying to be really nice to you because I value your friendship. You have been so special to me.” She kissed Javi’s cheek. “Is that okay?”

"Yes, that's okay," said Javi.

"Now," said Sarah, "you can kiss my cheek."

Javi started to, but then stopped.

"Don't you kiss on Algenia?" asked Sarah.

Javi shook kes head a little.

"Forgive me, please," said Sarah. "I'd like to introduce you to someone. Would you like to walk?"

"Yes, I enjoy walking very much."

They strolled up busy Main Street. Many people hugged and chatted, others stared into each other's eyes and nodded, and others walked, touching hands as they passed. One person got a surprise hug from behind and expressed an over-exaggerated sigh. A group was seated on blue and green mottled grass eating their meals. One of them finished and threw the plate against a tree. It broke into a hundred pieces and scattered on the ground. *That is so strange*, thought Sarah.

Up ahead, Sarah saw Zac, the supporter of General Voder, speaking with someone. Zac said, "Assertive. Yes, your vegetable production exceeds expectation, and there will be excess. You must leave it in the field to fertilize the next crop."

The farmer said, "No. No extra. I will harvest."

Zac raised his hands in disgust and walked away.

The farmer turned to see Javi and said, "Advice I restaurant maybe."

Javi nodded and said, "Yes, that is advisable."

The farmer waved goodbye to Sarah and Javi and moved on.

"Oh, no," Sarah whispered to Javi. "I have lost my sense of translation. That conversation made no sense."

"You are fine," said Javi. "That person must've said some words difficult for you to translate."

Sarah deduced Zac was providing his expertise on behalf of Voder to a farmer who didn't really want it. Javi had the voice of authority and sound advice. She felt proud about that.

As Sarah and Javi approached the farmland, Sarah could see up ahead a person and a tiger seated on the road.

"That's *my* tiger," said Sarah sadly.

"It is a beautiful creature," said Javi. "The person is enjoying communing with the tiger in a peaceful interaction. What a lovely site! But I'm sure it prefers your company more."

Sarah looked at Javi and chuckled. "You always know just what to say."

CHAPTER 10

CAVEMAN AND ANGRY ZAC

Later that day, Sarah walked up Main Street and saw two of the people who spoke in one-word sentences about to enter the Tall Grass Restaurant.

"Hello, my name is Sarah. How are you?"

"Hello. Caveman," one person said.

"Bull," the other said.

"Caveman?" asked Sarah. "Your name is Caveman?"

"Javi. Caveman. Name," she said.

"Oh, Javi named you Caveman?" asked Sarah.

"Yes," said Caveman. "Food."

"Yes, I am going to order, too," said Sarah, in slow deliberate words.

Caveman and Bull received plates of food and walked outside with their plates. Sarah took her plate outside and asked if she could join him. Cavemen nodded toward the ground. She sat down and tried to start up a conversation.

“How do you like living in Sedona?”

“Yes,” ke said.

“I think it’s beautiful,” said Sarah. “I love meeting new people and hearing about their home planets. Where are you from?” Sarah asked.

“Far.”

“Okay. I hope you are happy here. May I ask, do you have jobs?”

“Make. Plant. Tables,” Caveman replied.

“Oh, interesting,” responded Sarah, nodding appropriately. But she didn’t understand what plants had to do with tables. *They make tables. Does ‘plant’ mean making them out of wood? That’s possible. Or does plant mean making them at the manufacturing plant?* she wondered.

After a couple minutes, Caveman and Bull finished their meals and threw their plates up against a tree.

“Should I do that, too?” Sarah asked.

“Yes.”

She threw her dinner plate against the tree trunk, and it shattered and scattered on the ground.

“Why did we do that? Won’t that be hard for someone to clean up?” Sarah asked.

“Geophysical magnetronic disruption, GMDisruption,” Bull said.

Sarah was so surprised to hear him speak that way. "You speak in long words!" she exclaimed, her eyes wide with excitement, proud of herself for coaxing this out of him.

"Yes. You. Mean," Caveman said in a slightly angry, offended voice. She and a faint yellow aura and Bull stood up and walked quickly away.

"No!" she yelled. "I'm sorry! Please forgive me! Come back!" She shook her head in disgust at how rude she was with people she was trying to befriend. *How could I have done that?* she moaned.

She walked over to the tree. "GMDisruption," she recited. Looking down on the ground she saw tiny fragments of the plate growing smaller and smaller, slowly disappearing. *A self-cleaning planet,* she thought. *I like it.*

Sarah ran a couple miles to the golf course and asked if there was someone who could teach her and play a game with her.

"I have just the person. Me!" said a kind person who was polishing clubs and cleaning out the bags. "My name is Triol. What's yours?"

"I'm Sarah. I've only done this once before."

On Hole One, Triol demonstrated positioning the body and taking a swing. Sarah tried to duplicate the action, but hit the ball into the trees. "That's okay," said Triol. "Try again."

Sarah tried and tried, for an hour, but concluded that either Triol was not a very good teacher, or this course was not for beginners, or she didn't have the golf gene. "Do you have putt-putt?" she asked.

Triol laughed out loud. "What?"

"Putt-putt. A small course of only putting the balls into holes."

Triol chuckled, "No, I regret to say." Ke dared not laugh anymore for fear she may balk and run.

"How about ping pong?" asked Sarah.

With a shake of the head, Triol said, "No. Tennis."

"Softball?" she asked.

"Square ball," ke responded.

"Basketball?"

"Basket dump."

Sarah burst out laughing. "Basket dump! I'll have to see that one."

"They'll play tomorrow morning at the outdoor court."

Sarah thanked Triol and said, "Have a good day."

Triol waved goodbye. *I understand why Javi likes her*, ke thought.

Walking to town, Sarah saw Zak coming toward her.

"Smugly. Since you enjoy communing with deer, perhaps you should be an animal care-taker," he said, his eyes glaring at her.

"How do you know I don't have a job?" she asked, embarrassed to realize he knew she was seated amongst the deer and overheard his and Voder's conversation.

"Sternly. I know what every person does. You wander around a lot, cliff sitting. And…I see you want to be mayor." He lost his aura.

She looked down at her sandaled feet. "I'm not sure what I want to do yet. I used to be a teacher on Earth, but there are no children here to teach. This is really none of your concern."

"Condescending. If you need assistance, I'm sure I can help you find a job suitable to your abilities," said Zac, slowly and meanly.

"Perhaps I could teach you social skills" she said with attitude. She quickly said goodbye and walked away, Zak looking after her, shaking his head in disapproval.

Sarah decided to run the mountain, but only went partway up. She sat on a cliff and spotted mountain goats below her and some lizard-like creatures scampering among the rocks. She saw the little balls of fur rolling in the grassy park far below. A little rodent, something like a shaggy rabbit with beautiful calico markings, hopped in her lap. She petted the softness as she looked around, taking it all in, the entire panorama. Her vision telescoped to see distant farmlands, forested mountains and snow-capped mountains. Ahh, the heightened senses in her perfect body were more than she could ever ask for. She found it odd she could never see the sun. *Maybe the Brightness of the Beyond is just a dome with a big dimmer switch,* she mused.

Sarah rarely thought about her 26 years on Earth. Memories of fighting parents, loving grandparents, her first kitty cat, and falling in love were stored out of reach, like forgotten, old clothes in the back of a closet. She had new friends now, good friends she was getting to know well in just a few days. Spiritual, serene Javi. *Ke'd make a good Tibetan monk*, she thought. Aron the miller, singing and hugging his way through life, but he's also a mind-reading thief. And sweet, meek Mud who has some prejudices to address. All the people in Sedona were so nice, except the two she met from planet Frond. Had she already developed her own prejudice against people from that planet? No, it's not prejudice, she decided. It's just not liking their male dominance. That's all. She certainly liked Voder's musical ability. He was truly gifted. And Zac, well, Zac must have some redeeming traits—she just had to discover them. Frondians were hard to talk to, maybe because they announced their feeling first then continued with the rest of the sentences. And then there's Paul—still missing. *How often do people vanish?* she wondered. *It doesn't seem to bother the townspeople that they do.*

Sarah noticed a glow far away on the horizon to her left, a tiny pinpoint of light, and wondered if it was a big city, like Zion, described in the Bible. She could not see any other towns and wondered how far she’d have to walk or boing to visit one. Javi told her there were trillions and trillions of people on Resta. They must live somewhere.

Chapter 11

Townspeople

As Sarah met more people, she found herself busy all the time. She attended church worship services, went rafting, boating, and hiking, and attended baking competitions, animal seminars, and food factory tours. Some of her favorite activities were listening to philosophical discussions and going to concerts. Besides Voder's orchestra, there were other smaller ones and quartets from other planets. She loved them all.

One day, Sarah noticed the landscape had changed. A road undulated into the hills, and she saw skate boarders and cyclists really soaring! Once she noticed the river had swelled to capacity and it accommodated more sailboats. How can the landscape change so readily? Is this God at work? Maximum geophysical activity? Magic?

Sarah was mesmerized by many of the games people played. In capture games, semi-transparent fractal animals, about two feet tall, rose up out of the ground for a few seconds, were captured with a net, and taken to a pen. If not captured, they descended back into the ground, and rose again elsewhere, making some sort of pattern in the field. The challenge was to discover the pattern. Members of a team assisted each other by predicting where the animals would rise, then yelling and pointing to the ground. The team member closest to that spot attempted to capture the animal. There were two to four teams of four people each. This game required logic and athleticism as well as the technology to raise and lower the animals. Sarah watched a team captained by Zac slaughter other teams.

Sarah vowed to attend church every day, but occasionally missed a day—she was just that busy. She was happy to meet frequently with a group for meditation in the church's basement where there was a floor covered with soft fleece blankets. Everyone sat on the floor, Indian style, as Sarah would say, ankles crossed with their hands on their knees, palms up. Pastor walked around to each person and placed a hand on their heads and said a blessing. A sacred verse from the Book of Scriptures was read and everyone focused on it, thinking about it, tuning in all their senses as to what it meant.

After a few sessions, Sarah learned to stay focused longer and quiet herself completely, eliminating stray thoughts. In addition to her skills improving, she began to sense the depth of soul-searching in each person in the room. She could sense how close each person felt to God. But she couldn't understand why she should have these feelings: she didn't particularly like them and tried to tune them out.

At the end of one session, Pastor said to Sarah, "Don't worry about anything when you come here. Give your worries to Jesus. Let him speak to you and direct you."

Sarah nodded in agreement.

Then Pastor said, "I believe you are discovering your IED."

"My improvised explosive device?" exclaimed Sarah.

"No, no. Your ion exchange detector, IED. It's one of the most wonderful senses God has given us. It is used to detect a major change in the activity of someone's brain cells, signaling when a mind is focusing on the soul, expressing its desires and love. You can sense when someone is in a deep meditative state reflecting on the soul. I believe you called it 'depth of soul-searching'."

Sarah wondered how Pastor knew she thought of it as 'depth of soul-searching'. She decided not to ask. "I am beginning to understand my spiritual being more than I ever have before. It's truly wonderful. Pastor, do you really think I will make a good mayor?"

"Yes, you are the perfect counter to Voder's domination."

"Domination?" said Sarah.

"Voder's ridiculous lunacy, claiming that people need managing."

"I've never heard him say the word domination," said Sarah.

"That's how I translate it. He will never be a soulful person. I trust that you and I will work well together," said Pastor.

Sarah could not sleep that night. Was Sedona as pristine as she was led to believe? People are astoundingly happy. But are they truly as happy as they claim to be? Now, for a second time, Pastor was critical of Voder's soul. How can she say that? Doesn't God give everyone a soul? Perhaps Pastor is mistaking spiritual with something else. If you don't have to strive for excellence in your work, or build loving relationships, what's the point of living? She had overheard Voder ask that question. Zak's response was, "The point is to discover the point." Probably the point is to develop an old soul, as Javi had said.

She didn't understand everything Pastor said. *Does she read minds? And what's that about domination?* Sarah wondered. Still, she felt free and alive in a way she did not experience on Earth. When she looked at all the wonderful people of Sedona, she felt that her family and everyone she loved on Earth was having the same joyous, soulful experiences somewhere on Resta.

On a nature walk with sweet, meek Mud one afternoon, Sarah said how much she enjoyed meditation. Mud looked down and said, "I know you are wondering why I have a troubled soul during our meditation time."

"Oh, Mud, I'm sorry I picked up on that. I know it's none of my business, but I can't help it."

"It's okay," said Mud. "It's a God-given sense that we must use for good purposes. I think you are a good person. I may as well tell you: on Earth, I was a Pakistani girl, kept as a slave. I was abused and angry my entire life. Coming to Resta was the only good thing that ever happened to me, ever. And now I would rather be with birds than humans. Sad, isn't it?"

"I don't know what to say," said Sarah. "I'm so sorry, Mud. How old are you?"

"Here on Resta we don't have an age, there is no aging. But I was 13."

"Oh, no. I am sorry. I do hope you like being with me."

"I'm working on it," she said with her small, meek smile. "But I hate to say there are some people I will never be able to tolerate."

Sarah saw Mud's battered soul and vowed to encourage her in every way.

CHAPTER 12

HEAVEN

The nature group completed the hike and returned to town to find a large group of people gathered in front of the town hall. There was much ado about the establishment of a command center to oversee the operation of farms and factories.

“Why not?” yelled someone.

“There’s no point,” moaned someone else.

“It would ensure adequate food production all year, regardless of the season.”

“We don’t have seasons, and there already is adequate food production, for humans *and* animals.”

“But we don’t know what’s happening in other towns. We don’t know of their needs.”

“Food. More!”

There was more shouting.

“People do not need to be supervised or monitored of their activities! No bureau!”

"What is this about?" Sarah asked Mud.

"It has to do with Voder and Zac and some others," explained Mud. "They can't accept that Resta is a blessed place with God as our King. We don't need any general or prime minister. That's ridiculous!" Mud was stirred to anger. "I can't stand them!"

"Ah, the command center. It's okay," said Sarah. She put her arm around Mud and turned to lead her toward a distant restaurant.

"I'd rather go to church," said Mud. "Do you mind? I want to go alone."

"Are you sure?"

"Yes. I'll be okay," said Mud giving Sarah the tiniest of smiles.

Sarah watched Mud, head slumped down, walk toward the little white church. Then Sarah turned into a restaurant and sat at a small table in the back. She ordered a noodle dish and red wine. She thought about the day's events and how her idyllic life was changing. *People are arguing*, she thought sadly. *I don't like the people from Frond. Are most people kind only on the surface? Artificially? Javi is kind inside and out. I'm sure of it. Am I?* Sarah started thinking pastry chef would be a nice job. *I can't be mayor here.*

The townspeople decided the discussion should continue in a civilized fashion inside town hall. Someone formalized the topic—'Meeting to discuss the establishment of a farm management bureau'.

The notice nominating Sarah for mayor was removed. By the time Sarah arrived, the room was packed so she stood in the back with Javi. Various farmers voiced their opinions on the benefits of a bureau – ease of scheduling, food security, etc. Others were adamantly against it citing the dangers of establishing a hierarchy in a place such as Resta where love and compassion should dictate actions. The discussions went on and on and included cussing and crude comments. Voder, at the front of the hall, tried to lead the meeting speaking up in anger that people were being so closed-minded. Others began to show anger and frustration. Even Aron, whose planet became uninhabitable due to de-vegetation, became enraged that others thought that food should be the top priority. He insisted that relationships built on love are the reason people are on Resta and that God is the sole provider of food. Zac carefully explained that if anyone should command the center, it should be Voder.

As tensions continued to mount, several people spoke at once. The meeting was out of control. Sarah shook her head in disbelief and hooked her arm into Javi's.

"Stay calm," ke said.

"I can't." Then she shouted out, "Can I say something?"

The crowd calmed down, surprised to hear a voice from the back.

"I just want to know… **is this Heaven or isn't it?**" Sarah demanded forcefully. "**Well?**"

Silence.

“**Is this Heaven or isn’t it?**” she repeated, louder.

Voder spoke up first. “Disgust. Of course this is not Heaven. Resta is just our current home when our planets were destroyed. Everybody knows that.”

“But we have new bodies which may not age!” Sarah explained. “These could be our bodies forever! Our faces are all alike which may be the image of God! We understand each other’s languages! This is clearly a spiritual gift. The animals live in peace with one another! No prey, no predator. All of these things are a description of Heaven in Earth’s Bible. How can this **not** be Heaven?!”

Utter silence.

“Do you all know what Heaven is?” prodded Sarah.

Silence. Everyone’s eyes were wide, staring toward Sarah, but she could not tell if they were thinking about the validity of what she said or if they were contemplating her craziness. So she continued, “There is an abundance of food. All our needs are provided for! There is a brilliant light on the horizon which could be the Lord’s city.”

“Anger. I don’t believe it,” yelled Voder. “We are just people, individuals with human needs! We are not angels!”

“No,” said Sarah, “we are not angels. But we are not just people either. We are bodies with souls living together in one society, in community, in unity.”

"That is true," said Javi. "And this *could be* Heaven, or Heaven-like. God gives us everything, but we must strive to trust Him…"

"How can it be Heaven?" someone interrupted. "We have to work and prepare food and furniture and tools. God wants us to do a better job than we did on our home planets! We have been given a second chance to do it right! That means this is a place called Purgatory." This person's orange aura flared.

"No, this is the Promised Land," said someone.

"I do not know what that is but we do need to concentrate on our spiritual nature," said another person. When our planets were destroyed, we started over here, in our new bodies, in the New Garden, Utopia. And we especially do not need to be dominated by anyone."

"Dominate? We are not talking about domination! We just need better organization and a manager to run it. But not a general!"

Then Sarah shouted out, "Calm down! Let's solve this question. I am concerned that the word 'domination' has been introduced into this discussion accidentally. Manf from Lanet, I'd like to say a word to you and ask you to repeat that word back to me. OK? The word is 'command'."

Manf said, "Command."

"Good," said Sarah. "I translate the word as 'command'. Now here's one more word, Manf. 'Domination'."

Manf said, “Domination.”

“Good, I hear that as ‘domination’. Okay, Caveman from Far. Let’s do the same test. Please translate ‘command’.”

Caveman said, “Domination.”

Sarah said, “I translate that as ‘domination’. Here’s the last word: ‘domination’.”

Caveman said, “Domination.”

Sarah said, “I hear ‘domination’. Does everyone understand what has happened? The Far language, and perhaps others, mistranslate the word ‘command’ as ‘domination’. Perhaps this happens with other words as well. We need to clarify. One time I heard Zac say ‘command center.’ But we don’t really know what that means. Zac or Voder, please clarify.”

Voder said, “Assertion. An office where I and others provide expertise on crop production, grain distribution, storage, and issues that develop with people from other towns.”

“So we are talking about an office,” said Sarah. “Everyone shout back to me the word ‘office’.”

A few people said ‘office’. Gradually, they all said ‘office’.

Someone yelled, “Why do people disappear? Does f***ing Voder have something to do with that?”

"Wait! Calm down!" said Sarah. "No swearing, please!"

"Why not?" said the man in as angry a voice as Sarah had ever heard.

"Because it detracts from anything of *value* you have to say!" exclaimed Sarah, staring the man down. "Let's finish one subject before we move onto the next. The office that Voder just described sounds like the work Javi already does. I do not believe that Voder can do it better than Javi."

"Confidence. I will develop an overarching strategy to cover all aspects of crops, grains, distribution and neighboring towns," said Voder.

Someone said, "How dare you think you can command others and appoint yourself to such a position!"

Someone else said, "Maybe he is responsible for people disappearing!"

"Please stop!" yelled Sarah. "You are making me crazy! I swear we're almost done with this topic. We are not talking about commanding anyone. Are we, Voder?"

Voder looked at Sarah and slowly said, "Confidence. Let's say 'advise for the betterment of all'."

Javi finally spoke up. "Does anyone want Voder to provide agricultural-related advice instead of me?"

"No!"

"No, of course not." They seemed to all want Javi's expertise.

"Does anyone have any other significant new comment to add before we move to the next subject? Let's all think about this over the next few days before discussing it again. Now, about disappearances. Are people truly disappearing or just leaving?"

Someone said, "No one disappears from Heaven. That proves this cannot be Heaven!" A few laughed over the logic. Then everyone started yelling again.

"Everyone needs to work toward spiritual development. There should be no other job!" said someone.

"We don't need a mayor. A mayor will not help us learn to meditate or absolve our sins!" said Dora.

"A mayor will monitor who works at the manufacturing plant and explain the work that is done there," said someone.

"A mayor, maybe. But *not* a commander." And they were off to the races again.

Why can't they stay on topic? It's so obvious how they need to conduct their meetings. Why don't they see it? This rapid fire exchange of opinions made Sarah so frustrated she had to leave. *No, definitely not mayor. Not me.* Like a stone dropped from a cliff, her spirits sank, or rather, took a nosedive. *These people sound like typical angry mob Earthlings. Why are they so aggressive?*

Sarah ran to the library and pulled a book of Scripture from Lanet, Aron's home planet. She skimmed through the large book to find verses about Heaven, God the Father, and His son Jesus the Savior. Jesus' teachings included loving God with all of one's soul, loving one another, and abiding in the Spirit. She saw the word *Heaven* a couple of times.

Next she pulled a book of Scripture from the planet Frond. There was nothing about Jesus. It was all about God filling one's heart and being used for God's purpose. Did Voder believe he was fulfilling God's purpose? There was a general prayer similar to the Lord's Prayer from Earth, and it did include the word *Heaven*.

Next Sarah pulled the Psalm Book from the planet Algenia, Javi's planet. She found no references to Jesus or the Holy Spirit. It was all psalms to the One Great One, like in the Book of Psalms found in the Earth's Bible. *No Jesus?* thought Sarah. But she did see a reference to Heaven in which a body and soul can occupy the same space as another body and soul thus accomplishing a more intimate relationship with God.

There were other scripture books from other planets, but none as thick as Earth's Bible. She thumbed through a couple of those holy books. They mostly directed people to believe in and worship the One True God but many talked about the lesser gods. Sarah concluded most people in Sedona should have some concept of Heaven. Then why do they get so angry talking about it? And is there anyone who can figure out how to love Voder to bring some peace?

Sarah went to the little white church's meditation room and plopped herself down on the soft rug. The room was full of individuals, thirsting for calm or God's Spirit or a revelation: something to bring peace to their souls after the contentious town meeting. Sarah felt their yearnings. But she herself yearned for nothing. She wanted a peaceful emptiness so that God's Spirit could touch her soul, if He desired. She wanted to have a deeper faith in God—to know more about the Bible. She remembered having a new sense—IED, and oh, it was wonderful. She felt such love for these people who were searching for God, a depth of love she had never known before.

And then it happened. Sarah's mind emptied completely and she sensed a blazing, blinding white light that touched her deepest self, her soul. It was the first time she ever, in her entire 26 years, felt her soul. Then, in that bright light she saw souls and bodies occupying the same space as other souls and bodies. She saw, or somehow sensed, Jesus with His arms around Javi. They were smiling about the joy of Javi discovering that Jesus was the Son of God.

Suddenly, everything went dark.

CHAPTER 13

TUNNEL

Sarah plunged into darkness. She fell through the floor of the meditation room into an underground chamber. At least that's how it felt. She gasped in the total darkness, shaken at the jolt. Was this hell? Alone in the dark? She had just been experiencing awareness of her soul for the first time in her life, when this shocking event struck, and she laid in a heap and moaned. She actually hurt—her back. This was the first time she experienced pain or any discomfort since coming to Resta. Her new, perfect body was not so perfect.

She was able to stand up but bumped her head on a low ceiling. "Ow." She felt blood oozing out of a spot on her scalp, so she pressed on it with one hand, and, with the other hand, felt above her and all around. With that arm outstretched she took a couple steps in no particular direction until she felt a wall. She turned and continued walking slowly over the rocky floor, probing, and finding another wall, this one wet and slimy.

"Help!" she called out in her shaky voice. "Help, help!" All she heard in response was her echo. *Echo*, she thought. She remembered she now possessed the sense of echolocation. *How does it work?* She called out

Help and listened for the echo. She turned her ears out to widen the area she could hear accurately. She turned her body 90 degrees and called out *Help* again, continuing the process until she had gone full circle. She felt two opposite directions were more open, indicating she was not in a 4-walled chamber but a tunnel. She started walking, her back aching and head bruised, oozing. A large jagged rock hung from the ceiling, and she smashed right into it. Tears welled up in her eyes, but she kept moving, each foot exploring the floor before taking the step and her arms raised in defense.

A minute later she yelled *Help* again and decided she was going in a good direction. She soon discovered that saying *Help* one time gave her sonar-sensing echoridge all the information it needed to detect the direction of the tunnel. But after a few more minutes, she began feeling frustrated, the same feelings that arose when she couldn't find her parked car at the mall. Where was patience when she needed it most? Perhaps it was the darkness, the total blackness that made her feel lost and helpless. She stopped, knelt down and prayed for help. After a few more minutes, she heard a voice call her name.

"I'm here," Sarah yelled as loud as she could. Behind her in the tunnel, a flicker of light appeared. Then she saw Javi's aura.

"Oh, Javi! I'm here!" she called.

Ke came to her and put reassuring arms around her. "Are you alright?"

"Just a little bruised," she said. "What happened? What is this place?"

"I'm not sure," said Javi, "but perhaps you have found an old tunnel which used to be an escape route to other towns. It's mentioned in a book I read. I, though, assumed it was just rumor or legend."

"How did you find me?"

"I felt you disappear from the meditation room. Since you connected to your deepest soul, you were able to transcend space. I tried to sense your location, but all I could ascertain was that you were below the room. So I transcended downward."

"Transcend. I hate that word. I don't understand what it means," Sarah said.

"To go beyond the limits of," said Javi.

"How is it possible to transcend physical space or time? It's not! Even here our bodies are physical. Right? Our minds are derived from our physical brains, and our souls are… I don't know what souls are."

"Our souls are a gift from God, separate from our minds and bodies," said Javi. "But our souls are linked to our bodies. I always felt my soul was my deepest, innermost being, the only important thing about me. Let's get out of here."

"How? I don't know how to transcend. I can't."

"Me neither. Can you walk? I enjoy walking very much."

"Yes, I know you enjoy walking," Sarah said with a smile, her tears ending.

By the light of Javi's aura, they walked forward, not back to where they dropped into the tunnel because they didn't know exactly where that spot was. On they walked, and the tunnel enlarged slightly into a small cavern. Up against a wall were sacks of grain stacked neatly about 12-bags high, 6-bags wide and 2-bags deep.

"What is this?" asked Sarah. "Grain or flour?"

They examined the sacks. "It looks like grain," said Javi. "Why is it here?"

"I don't know," said Sarah. "I can't imagine why it would be here. Oh, look." She pointed to a frame on the wall behind the sacks. "It looks like a doorway."

They removed some of the sacks until there was enough room to squeeze behind them, and they wriggled through the doorway. They both stood there, stunned at the sight. In a large cavern were hundreds of wooden crates, some spilled of their contents—metal implements with handles, switches and barrels.

"Weapons," said Javi. "And here's the ammunition." Ke pointed toward a box of shells.

"Where did this come from?" asked Sarah.

"I don't know," replied Javi. "I have never heard of a stash of weapons. How can there be guns on Resta?"

"I don't understand," said Sarah in dismay.

They exited the large cavern and continued walking through the tunnel, slightly downhill, when they heard water flowing. They came to a narrowing in the tunnel with a river flowing through it. A passageway in the cavern wall took them up to a small opening lit from above. Fighting through the brush, they breathed a sigh of relief as they emerged into daylight. They crossed over several hills and found they were behind Mill Creek! They walked toward the mill. Aron was not there so they continued to the lane and on to Javi's home. Ke had a little vial of healing liquid and put a drop on Sarah's sore head and another one on her back. She instantly felt better.

"Thank you for rescuing me. You answered my prayer," she said.

"You are welcome. Let's find Aron. Perhaps he knows something about the tunnel and weapons."

They headed toward town and came across Zac first.

"Zac, may we talk?" asked Javi.

Zac didn't answer.

"Are you aware of someone hoarding grain away from the warehouse?" Javi asked.

"Irritation. No. But Voder and I have suspected this has occurred. We have not seen it nor know who is doing it. I am suspicious of the farmers behind Mill Creek because, for the acreage, they bring the least amount of grain to the mill."

"What would be the purpose of maintaining a stash?" asked Sarah.

Zac ignored her.

"Zac," asked Javi, "what's the purpose of maintaining a stash?"

"Weariness. Possible shortage in the future, or fear of an attack from another town. Excess grain could be used for bargaining. Why are you asking these questions?"

"Is this why you feel Voder should become the mayor or some sort of manager? To ensure proper grain storage and rationing if needed?" asked Javi.

"Irritation. Yes," replied Zac, "and if tensions mount between towns, Voder would be in a good position to lead militarily. He was an excellent General on Frond." He scowled and stared at Sarah to see her reaction.

"Zac, why do you think we are in danger? I personally don't see it," said Javi.

"Irritation. Because of the escape tunnels we have heard about. The previous civilization on Resta was a warring society."

"Tunnels to escape from what?" asked Javi.

"Weariness. Escape from one town to another. But they could easily be used for a stealthy advancement of soldiers. I have never seen the tunnels, but I suspect there is some truth to it."

"How do you know about the previous civilization?" asked Javi.

"Weariness. There are several books in the library that report the history of Resta. You may not understand, and the information is limited and incomplete."

"Thank you, Zac," said Javi. "You are most helpful."

"Mocking. Thank you, Zac," said Sarah. "You are barely helpful."

It took Javi a few seconds to understand Sarah's comment. Zac acted like he paid no attention.

They parted, both glad Zac did not ask if *they* found the tunnel.

They continued on, finding sing-songy Aron in a restaurant. He stood up and hugged them both.

"Please, have a seat," he said. "What can I do for you this beautiful evening, this lovely twilight?"

"We are curious what you know about Voder wanting to be a manager and military leader of this town."

"Just that," said Aron. "No more, no less."

"Are you aware of any towns that are expecting food shortages?" asked Sarah.

"No, none. None at all," answered Aron, thinking that was a strange question.

"Do you know of a secret cache of grain?"

"No, of course not. That would be ridiculous since we have a large warehouse. A huge warehouse filled to the brim!"

"I am concerned that Voder is pressuring you to submit to his management system," said Javi.

"Well, he tries his best, but I am unwavering in my response."

"I am glad," said Javi. "Thank you for your time."

Sarah said, "Aron, you have a good understanding of what Heaven is, don't you? I looked through the Scripture book from Lanet."

"Yes, abundance of food, image of God, understanding languages, but I didn't know we would not have our children," said Aron sadly.

Sarah could understand Aron's grief because he had so many children.

From that brief exchange, Aron detected that Javi and Sarah asked about the secret stash of grain because they had seen it or heard of it, and were very, very disturbed by it. But how? If they actually stumbled on the sacks of grain in the tunnel, they would not let it rest, but would pursue it endlessly. And if they discovered the armory… that could be disastrous.

From that brief exchange, Sarah and Javi sensed that Aron was telling the truth and did not put the sacks in the tunnel. They left feeling comfortable that Aron was a strong, trustworthy person who could stand up to Voder.

CHAPTER 14

BANA

The next morning, in her kitchen, Sarah said, "Javi, I guess you think I was foolish suggesting Sedona is Heaven."

Javi gave her an understanding look and said, "No, not at all. But what do you think now?"

"I feel naïve and wrong. I want it to be Heaven but it can't be. I feel that it has a purpose. What that purpose is, I don't know. After seeing two bodies pass through each other, I almost sense that bodies can become ephemeral. That would allow trillions of souls and bodies occupying the same space in the unity of the souls. I don't know what I'm talking about."

She changed subjects. "Whose weapons are those? The only one who might be interested in weapons is Voder. Don't you think?"

"I agree. Let's go ask him."

"Oh, great," said Sarah. "You just want to waltz up to him and ask why he has an arsenal."

Javi reflected for a moment. “By ‘great’ you don’t mean ‘large’ and by ‘waltz’ you don’t mean ‘dance’, therefore I conclude you are expressing sarcasm.”

“Yes, I am,” said Sarah. “I’m sorry. I don’t know why I do that. On Earth we always did it. It was almost…fun.”

“A societal norm,” said Javi. “I understand.”

“I do think it would be a mistake to ask Voder why he has an arsenal. That sounds too much like we are accusing him.”

“We could ask him casually if he knows of it,” said Javi.

“Maybe you should ask him,” said Sarah. “He may not appreciate me asking him.”

“Oh, okay,” said Javi slowly. “How could he have obtained them?”

“I don’t know,” said Sarah. “Perhaps they belonged to former inhabitants.”

“That is very likely.”

They thought for a moment.

“The only reason sacks would be stored there is to conceal the entrance to the other room, the armory. And I don’t understand why tunnels would be needed to escape from a town. I assume all the towns are like Sedona—free and open. Can’t we walk on the roads from one town to another anyway?”

"I agree with everything you say," said Javi. "But they could have been excavated by former inhabitants. Who knows what their society entailed? I wonder if there are people here today who arrived via a tunnel. Or maybe some Sedona residents left that way."

"Like Paul? I don't know. I feel that we need to find out so we can prevent it from happening again."

Javi nodded.

"I'd like to take a long walk—to another town," said Sarah. "Would you like to go with me?"

"No, I can't. I have an appointment at a farm today. Then I want to do research at the library."

"But I'm going to *walk*!" said Sarah.

"I heard. I am disappointed not to walk with you."

Sarah headed out alone, past the church, out of town, on to farmlands. The road was well maintained, even in the forests. She decided not to run or boing because she was hoping to spy unique birds and other animals, especially ones not seen on Earth. She walked for several hours when she noticed someone running speedily towards her. Sarah moved to the side so he could run past her, but the person stopped in front of her.

"Hello!" he said. "How are you?"

"I am fine. How are you? My name is Sarah."

"I am great. My name is Sanj, and I am pleased to meet you," he said.

"I am interested in visiting the next town down this road," said Sarah. "Is that where you have come from?"

"Yes, it is," Sanj said. "I am interested in finding a friend who lives in Sedona whose name is Bana."

"I don't know anyone with that name. Would you like me to help you with your search?"

"Yes, that would be nice. But I don't want to interfere with your visit to my town."

"It would be my pleasure to walk, or run, with you. And I'd actually appreciate hearing about your town," said Sarah.

As they ran towards Sedona, Sanj said, "We have a population of approximately five thousand. Several are from Earth. We have many recreational activities and opportunities to study nature and the Beyond. Of course, our food and wines are exquisite, a benefit of having a chef from Earth. Our pastor is a little lacking in spirituality, but don't tell anyone I said that," he said, chuckling. "He makes up for it in listening skills."

"Sounds very nice," said Sarah. "How far is it between towns? What's its name?"

"Hamlin. It's about a three-hour run. It is too far to walk in a day. Spending the night on the road is too dangerous. There are some animals that might bite you."

"I thought all the animals ate only plants."

"Oh, they do. But a large wolf may use your arm as a toy and shake it violently, for fun. I have heard of that before. Granted you may heal quickly, but it is still unpleasant."

"Did you know Bana from your home planet?"

"Well, yes!" said Sanj. "She is the daughter of my dear friend on Earth. I heard she lives in your town, so I am most anxious to see her again."

Lies, all lies. Sanj could not hint that Bana was his slave, given to him by her parents to pay off a debt. Now, he felt he could ask for her forgiveness and that all would be forgiven. Sanj then would be blessed by God.

"I am from Earth, too," said Sarah.

When they arrived in Sedona, Sarah and Sanj poked their heads into the church and some restaurants, but had no luck. They walked by the shops and around the park, but Bana was not to be found.

Down the road, Sarah saw Mud walking toward them. Mud looked up, spotted Sanj, made a squealing noise and ran away. Sanj did not seem to notice. Sarah thought it was curious, but didn't say anything.

Sanj felt confident they had sufficiently searched the town. He expressed some disappointment, but thanked Sarah profusely and headed back through town and off to Hamlin.

Sarah ran to Mud's house and knocked on the door. "Mud. Mud! Please open the door. Can I talk to you?"

But Mud's door did not open. Mad Mud. Sarah waited and sadly left.

As darkness fell, Aron walked from the park to Mud's house. Knocking on the door, he said, "Mud, it's me. Open the door."

The door opened and Aron quickly ran in. "Are you alright?" Aron asked.

"No, I'm shaking."

"Was that the man?" Aron asked.

"Yes, I can't believe he came here! Why won't he leave me alone?"

Aron put an arm around her and said, "Calm down. I'll take care of it. Don't worry. Shhh. Don't worry."

Mud took a deep breath to calm her nerves.

Chapter 15

Weapons

Javi spent the night reading through books that contained any mention of the history of Resta. Information was very scarce. There seemed to have been a previous civilization in which groups battled for manufacturing plants, farmland, and spiritual leaders who claimed to be God or gods. But there was no mention of how or when this civilization disappeared. Javi had no idea of the books' sources or accuracy. Perhaps it was all myth.

In the morning, Javi went to the huge manufacturing plant to see if weapons could possibly be manufactured there. Ke toured the facility and watched the production of waxed paper and bags for the bakeries, golf balls, waist cords for robes, composite wallboard for a new house, and an oven. The oven held a number of small metal parts—thermostat, knobs, some short wire contraptions, and racks. Javi decided the plant did have the capability of producing weapons. Most distressing. All of the manufacturing staff would know about it, but when ke asked them, they said they did not.

Javi sat in Sarah's kitchen as they discussed the previous day's events. Sarah spoke of the visitor she met on the road, Sanj, and how they couldn't find anyone named Bana.

"Is there anyone in this town with that name?" Sarah asked.

"Not that I've heard."

"Mud's reaction was so strange," she lamented. "I'm afraid she will never talk to me again."

Javi did not comment but finally said, "I actually don't understand what occurred. How would Mud know someone from another town?"

"I don't know," said Sarah, shaking her head. After a few moments, she said, "tell me of your discoveries."

Javi told her about the unsuccessful library research and possibility that the weapons could have been manufactured here in Sedona. But it seemed doubtful that such a secret could be kept. "Perhaps the manufacturing was done at night with a minimal crew. I, however, don't see a reason anyone would be willing to participate in this, unless there was a reward. And there are no rewards in Sedona. We are all rewarded every day with everything we need or could want."

"Let's say Voder is behind this," said Sarah. "He could have a few supporters in the manufacturing plant making the weapons, being promised rewards depending on battles fought and won."

"I suppose that is possible, but it seems like a stretch of the imagination."

"I know," said Sarah, exhaling. "It's a long stretch."

"Let's say the armory is left over from ancient wars. Voder could have found it, somehow, and tried to hide it behind the nearest available item—sacks of grain from the mill. He stole them and moved them into the tunnel. He's not interested in the grain itself at all. He just wants to hide the armory, thereby maintaining ownership of it."

"Yes, yes. I can see that," said Sarah. "But isn't the person most likely to discover the tunnel entrance Aron?"

"Oh, possibly. Yes, that would make more sense. But Aron knew nothing about a stash of grain."

"Javi, perhaps I didn't tell you that Aron lies. He told me," said Sarah.

"But I did not sense any falsehood during our talk."

"He is good at hiding it. He has experience doing this," said Sarah.

"I have heard of people from Earth lying, but never from Lanet. Was Aron lying when he told you he was lying?"

"Funny. I may never know that. What do we do now?"

"I propose we go back to the armory and examine the weapons. Let's see if we can determine when or where they were produced," suggested Javi.

They strolled down the lane toward Mill Creek, as nonchalant as people with no destination in mind could be. Luckily, the mill was not turning today. Seeing no one, they ducked into the forest behind, hiked over a few hills, found the opening and climbed down into the tunnel. The river water was fairly calm. They walked slowly and quietly, listening, in case someone else was there. But all was quiet. They removed some of the sacks and climbed into the doorway. They first inspected the open crates, looking for gun markings, signs of the metal composition, signs of wear, and the different types of weapons. Next they checked the closed crates for additional labeling. They discovered only two types of weapons, a short barrel and long barrel, but there were hundreds of each. Many crates looked as though they had been opened and nailed shut again. Some of the guns were scratched and dirty, uncared for. There were no words, letters, or identifying marks on the crates or guns themselves.

"I don't think Sedona produced these recently. They appear old and used to me," said Javi.

"If you say so," said Sarah. "Let's get out of here."

They replaced the sacks as neatly as they found them, and left.

****

Up on a cliff, Aron watched Javi and Sarah sneak around from behind the mill and onto the lane. His brown eyes narrowed to slits as he watched them walk along the main road towards town. Then he boing-ed down the mountain to Mud's house and gently knocked on the door. "It's me."

The door opened and Aron popped inside, closing the door quickly. "Avoid Javi and Sarah," he told Mud. "I saw them go to the tunnel. But don't be afraid. I'll take care of it."

"But you wanted her to be mayor," said Mud. "And now you think she's trouble?"

Aron felt a pang of guilt that he could make a mistake and cause her further distress.

"I know. I'll take care of it."

CHAPTER 16

HAMLIN

The next morning, Javi and Sarah made plans for the day.

"Let's visit that town," said Javi. "What's its name?"

"Hamlin, I think. Do we have something specific we are looking for?"

"I'd like to locate Sanj," said Javi "and ask if Bana used to live there."

"It's none of our business where Bana lives; that would be a weird question," said Sarah.

"But we can ask what life is like in Hamlin and visit some restaurants. Maybe ask if there is a mayor. Perhaps we can find out how frequently people travel and if others can attend their cultural events."

"Well, that does sound reasonable. Let's go."

They ran to Hamlin—an uneventful trip. Sarah did not enjoy the scenery as much as she was hoping to. Instead her mind was filled with worry about things that probably were none of her business.

After a few hours, the town gradually opened up in front of them. The multilevel houses were very large and the streets nearly empty. They reached the town center with shops and a theatre. Past that were hundreds of large houses, then farmlands and forests. There were only a handful of people walking on the streets.

"How can there be so few people outside? By the size and numbers of homes, there must be thousands of people living here," said Javi.

"Sanj said five thousand. Maybe people stay indoors," suggested Sarah. "I don't like this. This is creepy."

They walked toward a building that looked like the town hall and found a person inside.

"Hello!" he said. "Welcome to our town! My name is Clare. I am the Mayor of Hamlin. I am Mayor Clare, the last time I checked."

"Thank you," said Javi. "My name is Javi, and my companion is Sarah. We are from Sedona and are interested in learning about your town."

"Of course! What would you like to know?" asked Clare.

"What does the mayor's job entail?"

"Helping newcomers acclimate," Clare replied. "Ensuring housemates are well-situated and happy. Settling disputes. Keeping track of the residents: who arrives and who leaves."

"Your town is very large, but there are few people outside," said Javi.

"We prefer to stay indoors to enjoy the company of housemates. We meditate and discuss our spiritual natures. We find great reward in being soulful, examining our innermost thoughts and feelings and learning how to delicately approach others' feelings."

"Delicately approach others' feelings," repeated Javi. "What a wonderful manner in which to cherish and regard each other."

"How many people live in each house?" asked Sarah.

"Ten to twelve beautiful people who become a family."

"Having such a large town, do you have plenty of food?" asked Sarah.

"Yes, of course, God gives us an abundance," said Clare, giving Sarah a quirky look.

"I met one of your residents, Sanj, whom I'd like to see again," said Sarah. "Can you tell us where he lives?"

"Yes, I will take you to him," said Clare. "Does this have to do with Bana?"

"Yes!" said Sarah. "Sanj did want to see his dear friend, but we do not know of anyone with that name."

"Yes, Bana was a friend of mine, too."

"On Earth?" asked Sarah.

"No, here in Hamlin!" said Clare, doubling over in laughter. "A very unfortunate event… Such a loss…" he said, quickly straightening up.

"Bana used to live *here*?" interrupted Javi, pointing toward the ground.

"Well, yes. Of course. Next to Sanj's house. I'm surprised he didn't tell you. Bana ran off! Just ran away. We yearn for her terribly. I asked Sanj to go find Bana and convince her to return home," said Clare. "And then General Voder left… it was quite a loss…"

"Voder used to live here?" interrupted Sarah. She and Javi exchanged glances but tried not to look surprised.

"Yes, your town must be enriched by his presence," said Clare.

"Yes, of course," said Sarah. "Why did he leave?"

"He told me he desired to move, and I made note of it," replied Mayor Clare.

"Mayor Clare, we heard of two people who were in our town looking for other people from their planet. Did they come from here?"

"Well, yes they did."

"Was there no one from their planet here?" asked Sarah.

"No, I'm afraid they couldn't find anyone, so they left."

Sarah, Javi, and the Mayor arrived at Sanj's house and knocked on the door. Sanj answered and smiled.

"I'm glad to see you! What a surprise!" he said. "Come in! Come in!"

Sarah, Javi and Clare entered a spacious, well-lit living room. Beautiful, lush plants and flowers adorned the walls, and vines wound around the light fixtures and table legs. Well-spaced tables were arranged with four-to-six chairs. Scripture books and games lined bookcases. Several sofas lay against the walls and paintings of solar events and planets hung behind them.

They exchanged pleasantries for a bit. Then Sarah asked Mayor Clare, "How are housemates assigned?"

"We put only one person from each planet in a house. That encourages learning about other worlds and understanding their differences."

"Did Voder live here, too?

"Not in this house," replied Sanj. "But down the street."

"Do you know why Voder *wanted* to leave?" asked Javi.

"Yes," answered Clare. There was a pause.

"Can you tell us?" asked Javi.

"No. That is a personal matter," said Clare. "Thank you for understanding."

"Please accept my apology," said Javi with a slight bow. "We'll be leaving."

"I do have one more question, if you don't mind," said Sarah. "Have you heard of any tunnels connecting our towns?"

"Only rumors," said Clare. "I have never seen a tunnel."

"Thank you so much, all of you, for your hospitality," said Sarah. "Oh, just one more question. Is there someone here named Paul?"

"I'm afraid not," said Mayor Clare.

Javi and Sarah left Hamlin, unfulfilled by the short visit. On the run home they talked about the unexpected discovery that Voder also lived in Hamlin.

"Let's suppose," said Javi, "that he heard rumors of tunnels and weapons, and went hunting for them. He found a tunnel and traveled through it coming to the armory. He exited the tunnel, found the town of Sedona and decided to move there to possess and access the armory."

"That sounds reasonable," said Sarah.

"*Or* Voder knows nothing and decided to move only to find some musicians for the sole purpose of establishing an orchestra."

"That's probably more reasonable, but less exciting," said Sarah. "Now, on the other topic. Sanj says an Earthling named Bana ran away from Hamlin. We don't know where she is."

But," said Javi, "Mud sees Sanj, becomes fearful and hides in her house."

"And Mud won't talk to me because I'm the person who brought Sanj into town," said Sarah. "I hate thinking I am the one who has caused Mud such grief."

"No, you are not. Mud thinks of you fondly. I know it," said Javi in kes best reassuring tone. "She will calm down soon."

"What if… what if Mud *is* Bana?" suggested Sarah, her eyes wide with excitement. "She hates Sanj because she was kept as his slave. She comes to Sedona, changes her name, finds a vacant house and a job, and… is happy."

"His slave," said Javi thoughtfully. "That makes sense. We should leave her alone and pray for her happiness," said Javi.

"Yes, yes," nodded Sarah.

"Sanj, though, may continue his search," said Javi sadly. "I don't know what we can do about that."

****

Meanwhile, Aron decided to watch Javi and Sarah carefully to see if they made another trip to Hamlin or visit the armory. He also wanted to see if Sanj reappeared in town. So he got comfortable on his hidden perch up on the mountain and saw Javi and Sarah return from Hamlin, but Sanj was not with them. Aron decided it was crucial to block the tunnel entrance and destroy the armory so Javi and Sarah would stop snooping around. They could inadvertently lead Voder there. Revealing weapons to Voder could be disastrous. And no one needed to know that Mud used to live in Hamlin and traveled thru the tunnel to get away from Sanj. Resta needed to swallow the tunnel, make it disappear and stop all this inane curiosity. How could he accomplish this? He had to, to protect Mud. Aron sung to himself. *Oh Javi, Oh Sarah, what do you do and why do you care? Find other pursuits on this glorious day!*

CHAPTER 17

ARON'S TRIALS

Aron spent some time considering options for dealing with the armory. The weapons needed to be removed to a remote location, or destroyed, or… well that was all he could think of. Where could he move the weapons to? Was there a place underground? He could bury them. That would be back-breaking labor…unless he created a cave-in using explosives! Or what was the best way to destroy weapons? Crush them somehow and let GMDisruption absorb the fragments? Melt them into pools of metal? Think. Think. This was outside his field of knowledge. A giant explosion or a fire would upset the town. He needed to do this quietly. Perhaps a controlled explosion with the minimum amount of explosive necessary to do the job would be acceptable. He could ask for help from Triol, who used to be a miner on Algenia. But that would involve Triol in this nasty business. Crushing the weapons and letting them be absorbed seemed like a better idea. But how? Ahhh, the millstone! That may do it. He would do a test run with a few guns that night, after everyone went home.

Aron returned to the mill, praised the team for their hard work, and helped clean up. After they went home, he rushed to the tunnel and grabbed four guns and some ammunition to be crushed. He laid them in the path of the millstone. As the stone ran mightily over them, the shells popped, startling Aron. But it was not too loud—probably no one else heard. The guns made crunching noises. He stopped the stone by raising the lever from the river and inspected the damage. The guns were fragmented in perhaps five gnarled pieces each, not the degree he was hoping for. Either the stone was not heavy enough or the metal was too hard. At least the ammo crushed nicely so the pieces could be swept up. He'd have to do a thorough sweeping so pieces didn't contaminate the grain. He scattered the fragments of the guns and ran the stone over them again. Some of the fragments broke in half. He tried it several more times, but this was clearly not an efficient method. Aron scooped up the fragments and returned them to the armory.

Now what? Now what? He took a deep breath to ward off the panic he felt welling up inside. He prayed out loud, "Dear Lord, do you want me to destroy the armory and the tunnel? I feel that you do, but it's just a feeling. Do you want me to continue to care for Mud? I know you do. I will ask for help, because I can't do this by myself. Who would help me?" All he could think of was Triol.

****

In Hamlin, early in the morning, a musty farmer stumbled down a hole and discovered it was a tunnel—the tunnel. Ke was giddy to make the find, so ke clambered up out of the hole and raced to tell Mayor Clare. Clare and Sanj immediately set off on a trek in the tunnel, determined to find where it led and, hopefully, the whereabouts of Bana.

The climb through the tunnel was slow and treacherous, but they decided to persevere.

****

That morning, in Sedona, Aron went to Triol's house and asked for information about how much explosive material was needed to sink an area of ground about 800 square feet into earth that was undermined, like a sinkhole. This was Aron's ruse to avoid using the word tunnel.

"Well, it depends on the depth of the soil over the undermined area," said Triol. "If it's 5 feet thick you'll need 50 pounds of AX compound placed 3 feet into the ground. But if it's 10 feet thick, you'll need 75 pounds placed 6 to 7 feet into the ground. And it needs to be separated into 25 portions and placed in a grid fashion."

"I'll get back to you," said Aron, and he raced out the door.

"I can help you with the measurements!" Triol yelled after him. But Aron did not appear to hear.

How odd, thought Triol. *He is the strangest person from that galaxy I've ever met.*

Aron dashed to the tunnel, avoiding the mill, making sure he was not followed. Measuring the distance from the ceiling of the tunnel to the ground above proved impossible. The opening leading down was on a slant and very rocky. He thought a pipe on the ground pounded down, would appear through the cavern ceiling below, thus providing an accurate measurement. A 5- or 6-ft pipe would be good, but he was not tall enough to swing a sledge hammer onto it. A sturdy ladder would be helpful. He borrowed a step ladder and sledge hammer from a farmer and found a 6-ft pipe behind the manufacturing plant.

Aron pushed the pipe a few inches down into the soil in a spot corresponding to the ceiling of the armory, hopefully. He set up the ladder and climbed the steps with the heavy hammer. A simple pounding on the pipe sank it less than an inch. He did this a few times, but the pipe stopped sinking. It would require a full swing. He did a slow practice swing with both hands on the hammer, swinging its head from far behind his back, bringing it up over his own head and landing it surely on the pipe. That was good. He felt confident. So he made a full sturdy swing of the hammer, with some muscle behind it, missed the pipe and hit the ladder. It crumpled and fell over, sending Aron sprawling on the ground with a thud. The pipe popped up out of the ground like a leviathan spitting up a ship's mast. Aron laid there expressing a sigh of defeat and looked up at the sky, as the pipe fell next to him, and he wondered what to do next. "What, what, what," he chimed.

****

Caveman and Boulder, watching from the manufacturing plant roof three miles away, laughed and laughed over Aron's antics. They enjoyed the lunchtime entertainment.

Funny! Funny! Help?

Not help. Funny!

Yes. Help. Poor Aron.

Caveman and Boulder finished lunch and headed for Aron. But by the time they arrived, Aron had left.

****

Aron decided not to repeat his ill-conceived plan, so he returned the hammer and broken ladder to the farmer and went to see Triol. "Triol," he said, "I can't measure the depth."

"Let me help you," said Triol.

"No, no. The ground is too unstable. I don't want to endanger you, not you." Aron explained the pipe pounding that didn't work and asked for another suggestion.

"Well, you could try using a titanium-cobalt single chain magnet and a magneto device. The magnet is a stable clump of metal. The magneto device detects the magnet and will measure the distance between it and the magnet. Put the magnet in the undermined area and the device on top of the ground above it. Guestimate the height of the undermined area and subtract that from the distance between the magnet and the device. That will give you a pretty good idea how deep the soil is. Ah, wait. This won't work if there is too much lodestone or other magnetic substances in the soil."

"Where is the device and the magnet?" said Aron.

"There's a device at the plant and you'll have to ask about magnet. I'm not sure. Can't I come with you?"

"Thank you-you!" said Aron, and he went running out.

An hour later, Aron returned to the tunnel with the necessary items. He placed the magnet down in the cavern. He moved the device all around the ground, trying to find the spot below which the magnet lay, but there was no reading. Zero. He took the magneto device down into the tunnel and approached the magnet. He saw the needle start to register, but it was not stable. It flopped up and down. "Arghhh, arghh!" he yelled in frustration. There must be iron compounds all around. He returned the items to the plant and went to see Triol.

"Triol! There must be iron everywhere! It didn't work!" Aron was not singing now.

Triol was slightly amused by Aron's frustration, but decided it would be kinder to appear dismayed. "Please!" Triol said, "Let me come see this place. I can't believe your problem is unsolvable."

"Triol… Triol," said Aron, wiping his face and head in his hands. "I need your help, but I don't want you involved."

"Involved in what?" asked Triol.

"This nasty business. I found something I wish I never saw. And now I want to destroy it. But I don't know how. So I thought I could bury it with a cave-in, but I can't do it alone, and now I don't know if it can be done at all…"

"Tell me what you found," said Triol.

"I may not have much time."

"Take me with you. Maybe I can help, somehow. Please," said Triol.

Aron looked into Triol's eyes and mind. Triol then understood Aron's urgency and the seriousness of the situation. Ke nodded and the two of them ran out toward the lane. They ducked into the forest toward the tunnel. When they stopped, Triol saw no sinkhole. Aron looked around to make sure they weren't followed and took Triol to the tunnel entrance, and they climbed in.

"So the stories are true!" Triol said. "This is the mysterious tunnel." To Triol's further surprise, a large cavern opened up and he spotted the sacks. Aron moved some of them, exposing the door. They climbed through the doorway, and Triol was truly shocked at the sight. "What is all this?" Ke picked up a gun and examined it. "How can this be?"

"I don't know," said Aron. "All I know is we've got to get rid of it before someone finds it."

"Who?" asked Triol.

"I can't tell you."

"Okay. How much time do we have?"

"I don't know…exactly."

"So we can bury them, warp them, crush them, melt them…," said Triol.

"Yes! I tried crushing them with the millstone but they only broke into a few large pieces. And the bullets popped loudly when I crushed them. Let's melt them."

"I don't know how to do that on such a large scale. It would be difficult even on a small scale," said Triol. "Let's break them, then bury them. It may be faster than trying to figure out how to crush them enough for GMDisruption. If they are simply fragmented once or warped slightly, they are un-usable. Then, if they were ever dug up in the future, they'd still be un-usable."

"Triol," said Aron with relief in his voice, "you've got it! Can we use the millstone?"

"How many pieces did a gun break into with one pass of the stone?" asked Triol.

"About five."

"That's more than adequate. How many guns can we crush with one pass?"

"I don't know. Perhaps 50?" guessed Aron.

"That's great. Now, where can we bury them? How about where they are now?"

"Yes, that would be the fastest. Can you help me?"

"Yes, Aron. Of course. I am delighted to help. Thank you for confiding in me. I understand why you want to keep your actions concealed. You can trust me."

Aron held up his green-aura'd hand to touch Triol's. Triol's blue aura shimmered brightly. They exchanged thoughts and developed a blur between their bodies, establishing a bond of trust they never shared before. They knew it would require their bodies and souls to accomplish this feat.

CHAPTER 18

DESTRUCTION AND RESCUE

That night, after the mill workers left, Aron and Triol stole away to the armory, grabbed a crate of guns and hauled it to the mill. They placed the guns in the path of the millstone and set it rolling. Crunch, crunch. After two passes, they used a flat-bladed shovel to scoop up the fragments and dump them back into the crate. They ran it back to the armory and grabbed a second crate. Repeating the procedure 200 times took all night; but, they finished destroying all the guns. Aron was nervous about crushing the ammunition so they followed Triol's suggestion of soaking the rounds in oil first, then crushing them. No sparks flew and no pops could be heard.

When the Brightness of the Beyond started to brighten, and the scenery came to life, Aron and Triol gave the mill one final sweep and went their separate ways. Aron breathed a sigh of relief, so grateful for Triol's assistance but wondering if ke could keep a secret. Probably not. But that was a worry for another day.

Triol began to devise a plan to collapse the armory ceiling in stages. Ke collected the necessary items from storage bins in sheds behind the warehouse. Ke knew where these materials were located from having to flatten some hills on the golf course. Ke would use a minimal amount of explosive and gauge the response, increasing the amount until the cavern caved in.

That evening, Aron and Triol began their second covert operation. They dug a few small pits in the ground and set off some charges. From inside the tunnel they could see clumps of dirt and rock had fallen from the ceiling near the bags of grain. They adjusted the number and location of the pits and set off more explosives. After two more blasts, the rocky ceiling finally crumpled on the crates of fragmented guns and ammo, and the whole ground above collapsed.

They hid the tunnel entrance as best they could by shoveling some dirt around and covered it with branches. At one point, they felt a tremor, a tiny shaking of the ground. A small neglected charge exploding? Probably not. Maybe a large crashing of rocks in the tunnel? Possible. Or perhaps the river following a new path around fallen rock. But the tremor dissipated quickly, so they gave it no additional thought.

When all was done, Aron and Triol looked at the plot of ground they had destroyed.

“This looks terribly horrible!” said Aron. “I hope this mess will discourage people from rooting around down there.”

“Maybe greens will colonize the area quickly,” suggested Triol. Greens were small woody plants that grew rather quickly into four-foot high shrubs with pink or purple flowers. They were found everywhere so would not pinpoint the location of anything out of the ordinary.

Suddenly, the ground rumbled loudly. The monstrous Resta was awakening. Aron and Triol could hear it and feel it vibrate their brains and teeth.

"What's happening?" cried Aron.

"I don't know," said Triol. "Is it the river?"

They looked down at their feet and ran back up the hill heading toward the mill. Then they turned, and, with horror, watched water explode out of its underground bed and gush across the cave-in site into the tunnel.

"How do we stop it?" Aron was filled with panic and couldn't think.

"I don't know! What can we do?"

"I don't know! Maybe we don't want to do anything!"

"That's not a solution!"

They waited there for what seemed like many tense hours. Yes, the water was slowing down. The noise of the raging river toned down a bit, and they could imagine it would eventually stop. After another eternity, it diminished to a slow pooling then drained somewhere in the mud. They could see that the armory was filled deep with rocks and mud, and that dispersed along the tunnel, were the sacks of grain, muddy and unidentifiable.

“Well, this is the final product,” said Aron. “The question is: Were we successful?”

“Maybe. We’ll have to see it in the daylight.”

“Oh, Lord. What a terrible mess! What a horrible, horrible mess!” lamented Aron. He still could not sing.

“It’s alright,” said Triol. “It’s the best we could do.”

****

The long, slow hike in the rocky tunnel was difficult. After many hours Mayor Clare and Sanj began thinking this was not a possible route away from Hamlin to Sedona. Gradually, they became aware of a rumbling noise, a distant vibration they could feel through their feet. Ever so slowly, fear crept through their bodies, and they grabbed onto each other’s shoulders.

“Go back! Go back!” yelled Clare.

They turned and took off running as fast as they could which was futile due to the rocky ground and the head-banging rocks above. They slowed, and their fear intensified as the rumbling grew louder and louder. Suddenly the noise turned into rushing water crashing through the tunnel, pounding on their bodies, sweeping them off their feet. They were banged up-down-left-right and every which way. Clare was sure he had a concussion and would die. Sanj wondered how long he could hold his breath.

****

Meanwhile, Javi and Sarah sat serenely in the meditation room below the church, in a calming meditative state that revealed their joyous souls. Unexpectedly they felt a faint ground tremor and became aware that a panic situation was nearby, some vibrations mixed with fear. Or was it anxiety? No, it was truly fear. They were cast downward, physically downward, soul and body, into the tunnel.

Not again! thought Sarah.

But this time, they transcended into water, fiercely rushing water!

Javi grabbed Sarah's arm and wedged kes feet against rocks on the tunnel floor to keep them from getting swept away. Ke looked at her with calm eyes. Sarah trusted Javi completely, those eyes and that aura.

Suddenly, something crashed into Sarah—a limp body, a human body. Sarah grabbed the body and closed her eyes focusing on her deepest self, willing a transcendence upward into the meditation room. She felt she was being poured out of a 50-gallon drum, tumbling over onto the fleece rugs, still holding onto an equally drenched victim. She wiped her face and her hair back and saw Mayor Clare whom she tried to sit up, but it was all dead weight.

Meanwhile, Javi appeared holding onto Sanj, in seated positions, coughing up water. “Thank you, thank you!” cried Sanj, gasping.

Javi looked over at Sarah and smiled.

Pastor ran to get the healing oil and put a drop on Mayor Clare’s chest and one on his head. After just a few moments, Mayor Clare started coming out of his traumatized state: eyes focusing, trying to speak.

Javi said to Sarah, “You did it. You transcended out of a dangerous situation, and you carried a person with you. Your spirit is strong. Your body and soul are working together. God has given this gift to you.”

Javi hugged her, and she thought, *thank you for teaching me. You are my soulmate.* She was glad ke could not hear her thoughts. She happened to glance at her own hands and saw, for the first time, a faint purplish glow. A haze really. Sarah knew what it was—an aura. Her own aura appearing for the first time. She was ecstatic!

Mayor Clare began babbling, “I was so afraid! I’ve never been that afraid in my life!” He looked at Javi and Sarah. “Thank you for saving me. How *did* you do it? That was so scary. Oh, Sanj! I forgot you. Are you alright?”

“Yes, I’m fine. It’s a miracle that we are here.”

“What kind of a tunnel is that?” quipped Clare. “I’m never going down there again.”

“Me neither!” said Sanj.

Pastor found some dry robes for them to change into and brought down some tea and cake.

"Why were you in the tunnel?" Sarah asked Clare and Sanj.

"We are interested in finding Bana, of knowing if this was a suitable route for traveling," explained Clare. "I was not expecting a flood."

"Of course not," said Javi. "We knew the tunnel was there, but not why it would be flooded."

Mayor Clare and Sanj thanked them again and left.

Sarah stuffed a piece of chocolate cake in her mouth. "We need to go back to the mill and see what happened."

Javi nodded, and they left.

Mayor Clare and Sanj ran toward Hamlin. They were watched by Aron and sweet, meek Mud who sat up on the cliff. Mud cringed at seeing those people here in Sedona. She felt hunted in a search that would never end. Aron felt her tremble.

"Just ignore them. It will be alright," said Aron gently.

"You keep saying that, but it's not alright," said Mud, angrily. "I can't stand the thought of seeing him. I start shaking just thinking about it! And if he talks to me I will die! I know I will just die!"

Then they saw Javi and Sarah exit the church.

"It will be alright," Aron said again gently. "I will make it alright." He wanted to be the father she never had. He was not a father to his own children anymore, but he felt so grateful to have Mud. He would protect her and let her become truly free. She deserved to have Sedona be her utopia, her paradise.

Chapter 19

Rumbling

Voder and Zac were in the park when they heard the rumbling noise. Voder was enthusiastically reminded of distant rocket-propelled grenade explosions back on his home planet, but the odds of such a thing here was zilch. Ah, he missed those days. Zac thought it sounded like a fierce waterfall. Echolocation suggested the hills behind the mill as the origin, but seeing no one, they trotted toward the manufacturing plant. Then they spotted Caveman and Boulder on the roof of the plant looking toward the hilly terrain behind the mill. Voder and Zac sprinted and hopped in that direction over the bushes and rocky outcrops. Jumping up into a tree, they scouted for anything unusual, not knowing what they were looking for.

"Confidence. There," said Voder, pointing to a spot where the edge of a pasture was sunken and muddy, and water was gushing up out of the ground. They jumped out of the tree and bounded toward the pit.

Many people felt the tremors and headed for the mill. Searching across the landscape, they saw Voder and Zac in the hills and trotted along a course to intercept them. A continuous line of people, including Sarah and Javi, streamed from the town to the mill and toward the cave-in. They saw the enormous crater, muddy and rocky. The underground river was shooting up at one end of the crater and then seeping down into the muddy bottom. Then the rumbling died down.

Dora, one of the mill workers, yelled to Voder, "What's happening here? What have you done?"

"Rage. I have done nothing," replied Voder. "How dare you ask *me*!"

Dora directed the next question to everyone. "Who knows what happened? Anyone, anyone?"

No one answered.

Dora continued, "It appears that a sinkhole or cave-in has affected the underground river. If that flow has been diverted, it will affect the mill. We rely on that river to turn the millstone. So what happened?"

Still no one answered.

Dora moved through the crowd, gently touching each person's arm or shoulder. She sensed anxiety and fear that Resta's ground was not as stable as expected. Everyone's home planets had very active geology. But Resta? Resta should be sound. Dora touched Zac's arm and sensed some confusion: that was all. Then, she reached Voder and put a hand out slowly to attempt a gentle passing of fingers.

Voder suddenly took a step back while his aura changed to dense brown, his eyes narrowed to slits, and his eyebrows furrowed. "Anger.

I do not know what happened," he said sternly. "You should trust me above all others. You Lanet freak."

"What?" spurted out Dora. "You have mistaken me for a beautiful person from Lanet. But I am not. I am an empath from Narth. I have no reason to trust you above anybody else. Not when you have secrets. If you know something that affects this town, you need to tell us and tell us now!"

"Indignant. A female has no right to tell me what to do," yelled Voder, feeling the power of self-righteousness.

"I am not a female. I'm a person equal to you, Person!" yelled Dora.

Voder's appearance softened somewhat. He hated everyone who was not male. He remembered now his attitude was wrong on Resta, but it was so hard to repress. He had no allies in Sedona except Zac. It was time for him to leave. He wished he never came to Resta. He suddenly felt weak, his knees about to buckle. He stumbled away.

Aron stood at a distance, closer to the mill, shaking his head, wishing he had planned this better.

Ever so gradually, the low rumbling recommenced, like a machine starting far away. Then it grew louder and could be felt through everyone's feet and up to the knees. They looked at the ground and all around.

"What's that?" someone asked. "What's that?"

The mud in the pit made gurgling noises and started bubbling up. The churning grew wilder as splintered crates were pushed to the surface.

People started moving up the hill to distance themselves from the pit; but, they turned to see the strange event.

“What is that?’ someone asked. “Metal?”

“Chunks of metal,” said someone else.

“And there is a cylinder shape. Those are guns! Pieces of guns!”

Everyone gasped over the horrible realization of weapons on Resta. But this planet, in its beauty, had a process for sending the guns deeper into the earth. GMDisruption did its job. Gradually, the debris disappeared, and silty soil deposited evenly throughout the pit. The soil level slowly rose until it reached its original height, level with the ground. The rumbling stopped, and all was quiet.

Pastor spoke first. “Thank God for this blessing upon us.”

Quite unexpectedly, the low rumbling could be felt again. Everyone looked down at the ground and took a few steps back. Mud screeched as she saw the dirt under her own feet gurgle. She sank a few inches and tried to jump out, but her feet were suctioned firmly in place.

“Aron! Aron!” she yelled.

He ran down to her, grabbed her and pulled with all his might to yank her out. Javi ran down from the top of the hill to assist. Sarah started to follow Javi, but hesitated out of fear, then puffed up her courage and continued. The three of them, with all arms wrapped around Mud, yanked hard.

****

Meanwhile, in the basement of the manufacturing plant, Caveman and Bull sat in the technology center looking at a monitor which showed an aerial view of the hills behind the mill. The screen showed GMDisruption activity. Bull turned knobs to increase intensity as requested by Triol and Aron so that the guns and crates were fully absorbed. Red dots on the screen showed debris, and green dots indicated humans. One red dot stood apart from the rest. Wondering why this one piece of metal was isolated from the rest, Bull zoomed in to see that a green dot was under the red one. With a fearful gasp, he suddenly hit the STOP button.

****

Mud was instantly free, not because of her friends' effort but because the suction stopped. The ground had finished its reclamation of debris, restoring Resta to its pristine state.

Mud breathed loudly and held onto Aron.

"You are safe now, safely in my arms, my child," he said.

Sarah and Javi were terribly relieved Mud was safe and acknowledged Aron's affectionate, parental tone.

"Let's go back to town," said Javi.

The four of them walked slowly up the lane and along Main Street.

Aron started in a sing-songy speech, "I don't know how to tell you, to tell you, how much I appreciate your help, help, help." He looked toward Javi and Sarah. "I can't do this alone: hold a secret, forever alone, alone."

Mud's emotions started to well up. "Thank you for helping me," she said to Sarah and Javi. "I am grateful that you care about me."

"Of course we care about you," said Sarah. "Very, very much. I don't understand why you were sinking. It doesn't make any sense. But don't worry. We'll figure it out. Okay? Don't worry."

People were in the habit of telling Mud not to worry about stuff when they really didn't know what needed worrying about. She closed her eyes, not wanting anyone to see her guilt or sinful thoughts. *I am not sinful*, she corrected herself. She was glad Aron did not know what she was doing. He had not read her mind in a long, long time. She let Aron lead her as they came upon the crowd gathered by the town hall.

Pastor and Dora were talking. "Where did the guns come from?" asked Dora.

"I don't know," replied Pastor with her head shaking in disbelief. "How would I know?"

"It appears someone knew about the weapons and tried to bury them," said Dora.

"Who knew?" she asked of the crowd.

"I did," said Aron, working his way to the front of the crowd.

"No!" gasped Dora. "My dearest friend? How…"

"Yes," repeated Aron. Then he whispered to Mud, "I did it for you."

"No," whispered Mud, turned toward her savior.

"I must," said Aron, not looking at Mud. "I have to set the record straight." He stared right over the crowd.

"I'm sorry," whispered Mud with tears in her eyes.

"I must," repeated Aron, looking toward the crowd who began wondering what was wrong with him. "I found the weapons, and I buried them so no one else would find them. No one else."

Anger rumbled up in everyone that Aron had done this in secret. Aron! A respected and important friend to all had the audacity to take this on himself. Did he desire to raise his own stature? Did he even think about the river flow? Being the miller, he should've known better.

Sarah did not know whether to speak up or hide the fact that she also knew about the guns. She really did not want to talk about the tunnel, but thought it best to be honest. Upon opening her mouth to speak, Aron moved his hand to touch hers. *No,* he thought. *Do not say anything. It will not help.*

Sarah agreed in thought. She transferred the message to Javi by slowly shaking her head *no*.

CHAPTER 20

VODER'S TRUTH

Later that morning, Sarah, Javi, and Aron decided to return to the tunnel to inspect the entrance. From a safe distance behind, Voder and Zac followed them, remaining hidden but close enough to hear.

Javi, Sarah and Aron found what should have been the tunnel entrance, but it was filled with rocks of all sizes.

"Aron," said Sarah. "The explosion sent water through the tunnel and nearly killed two people from Hamlin. Did someone help you with the chemicals?"

Aron did not reply. He could've lied to Sarah telling her he laid the explosives himself. But he felt Sarah deserved better than that.

Just then Javi said, "It was Triol. Wasn't it?"

"Yes," replied Aron. "But please keep that to yourselves. I don't want anyone to know."

"Excited. So this tunnel leads to Hamlin," declared Voder, coming down the hill.

"Yes, it does," said Javi. "I have traversed it, or nearly the length of it."

"Query. Anyone else?" asked Voder.

"Mayor Clare and Sanj," said Javi.

"Query. What were they doing here?" demanded Voder.

"They were looking for someone," said Javi.

"Anger. Why didn't you tell us about the guns?" asked Zac.

"Because…" started Sarah. But she really didn't know what to say. Could she admit that she didn't trust everyone? "Because I felt the fewer people that knew about them, the better off we'd all be."

"Rage. How can we trust *you*?" demanded Voder, his wavy black aura becoming sharper.

"I don't have any reason to use them, own them, distribute them, or…or…anything."

Sarah was floundering here and shook her head over her inarticulate response.

"Friends," said Javi, "we tried to determine who knew about them and from there, ascertain their potential danger. I am grateful to Aron for destroying the arsenal. We should all be grateful the incident is over."

“There could be additional guns somewhere,” said Aron. “The only person I know who wants a military is you.” Aron pointed at Voder. “You. You!”

“Anger. I know nothing of guns. I have no interest in warfare.” Voder’s sharp aura disappeared completely.

That was the most ludicrous thing Sarah heard all day. All month. Everyone knew *that* wasn’t true.

“Didn’t you use to live in Hamlin?” asked Aron.

“Confident. No,” said Voder after a pause.

“Mayor Clare told us you did,” said Sarah.

“Maybe you and Mayor Clare knew about the guns and were planning something,” said Aron in an angry, accusing voice.

“Frustration. That idiot. I moved here, but it had nothing to do with guns.” Voder’s face changed to one of anguish. “No. I had to leave Hamlin. I couldn’t bear it.”

“Bear what?” asked Sarah.

“Regret. That’s none of your business.”

Aron took a step closer to Voder, but he backed away.

So Sarah took a step closer and touched his arm, feeling it tremble. “What is it?” she asked so gently.

"Sadness. Being so close to Kara—my baby," he said, dropping his head and moaning. "I couldn't stand it. It was too painful, seeing my grown baby."

"Your baby!? Your daughter?" asked Sarah. "You should be ecstatic that you and your daughter appeared in the same place. What are the odds? Clearly this is God's plan for you."

"Guilt. But I fear Kara will learn the truth…" And Voder could speak no more.

So Aron read Voder's mind, and Voder tried, but could not block it. Aron revealed, "That you didn't really love her—your baby girl. Not in the same way you loved your son."

"You have the chance to love her *now*," said Sarah. "What luck! Don't pass this up!"

"Guilt. It hurts too much," said Voder sadly, his eyes filled with tears. "I need to leave Resta."

That left everyone speechless. No one *wanted* to leave Resta.

"Voder, perhaps you can't leave Resta. No one can. Why didn't you love your baby?" asked Javi.

Voder couldn't form the words, so Zac spoke up. "Prideful. Because on Frond, females were not highly regarded. Sons were prized for their physical abilities, strength, and intellect. For their value in society."

"Regret. My first child was a son," said Voder. "My second was a sickly girl who died young. When I decided to try again, I wanted another son. But instead, Kara was born and I never paid attention to her. I'm afraid that in her soul, she knows of my practices."

Sarah stood there aghast. "I am not enthusiastic about the society you lived in, but that was your culture. And now, you want to *continue* that culture by ignoring your daughter?" Sarah threw up her arms. "Do you screw up on every planet you go to? I am totally astounded!"

Looking at Voder, Zac said, "Consolation. I was wrong, Voder. Your purpose is not to command an army or a town. Your second chance is to love your daughter. Go to her."

"Humbly. You're right. You're right. I will beg her forgiveness," said Voder.

"Voder, has Kara actually seen you and knows who you are?" asked Javi.

"Guilt. I don't know," said Voder.

"Then you don't know what her reaction will be," said Javi. "She may run to you with open arms. She may sense your grief and comfort you. You cannot assume she will be angry. You may be a complete stranger to her. You need to meet her."

"Compassion. Please, please," said Zac. "Go to her. I will accompany you, if you like."

"Dejection. My behavior… It's pathetic," said Voder wiping his eyes, looking ashamed. "Alright. Yes, let's go."

Voder and Zac walked to the road. They picked up the pace, and ran out of town.

Sarah tried to offer an explanation. "Here's something we didn't know about the society on Frond. That they ignored their young females. That's not that unusual. There were many societies on Earth with the same characteristic. We do know this. We should've realized something like this was happening to Voder. We could've helped him a lot sooner."

"Do you think they'll be back?" asked Aron.

"No," replied Sarah. "They will find they are richly blessed by staying in Hamlin. But we should invite Voder back to be a guest conductor sometime."

"That's a good idea," said Javi.

"Must we?" said Aron.

CHAPTER 21

GUN AND MEETING

The restaurants were filled to capacity, and everyone was chattering about the events that morning: the presence and destruction of the armory and Voder's return to Hamlin. Rumors ran rampant about who besides Aron knew about the weapons and where they came from. One group of gossipers said they heard that the armory was from a past civilization and that someone from that civilization could come to retrieve them. So it was necessary to search everywhere to confirm there were no additional weapons.

"Who said that?" said someone. "We have no proof of any of that."

They decided a town meeting in the afternoon would be in order. Someone posted a notice on the town hall door, and it was requested that the Pastor conduct the meeting. The Pastor agreed on one condition—that Sarah assist. Sarah was surprised and unenthusiastically said okay.

Sarah and Javi brunched together then strolled down to the church, but it was packed. So they took the stairs down to the meditation room. They emptied their minds of all their thoughts and worries so that the Spirit of God could fill them. In sparkling clarity, they sensed the love in people's hearts and the earnest desire to be a blessing to each other. Why couldn't this love make an appearance during the town meetings?

Sarah prayed for Voder's and Kara's relationship. Javi prayed specifically that Voder's chauvinistic ways would be forgiven and that he could break down that wall in his character that was so strong and admired on Frond and be relieved of the grief that chained him. Sarah sensed what Javi was praying for and thanked God that kes spirit had such depth and love.

On the way to town hall, Sarah said, "We figured Voder all wrong."

"Does that mean we misunderstood him?" asked Javi.

"Yes, that's a better way to phrase it," Sarah said. "He left Hamlin because of his daughter, not to gain the arsenal."

"Do you think that Pastor told him he had to leave Sedona?" asked Javi.

"Maybe," said Sarah, "but I'm sure he would've completely ignored her. Maybe it's one of the things they argued about."

She looked up and saw Mayor Clare and Sanj from Hamlin standing in front of the town hall.

"Why are they here, again?" whispered Javi in disbelief.

"This is going to be a mess," said Sarah.

They approached the visitors, expressed cordial greetings, and suggested they all move away from the door to have a private conversation.

Aron, at the doorway, recognized Sanj and grew anxious. What if Mud walked down the street? Sanj would see the missing Bana, the girl he had tortured in another life. Sanj was her master and head of household where he enslaved and abused her. Aron prayed Mud was not walking toward town hall.

To Javi and Sarah, Mayor Clare said, "We are concerned over what we have heard, that there was an armory here. We will not stand for the militarization of Sedona!"

"I assure you," said Javi, "the weapons have been destroyed. They were in the tunnel which is now probably unpassable, but it is too dangerous to verify."

"Thank you," said Mayor Clare. "On a different subject, we are overjoyed that Voder has returned to us. The mended relationship will bless the entire town and bring joy to us all."

"We are pleased, too," said Javi, winking at Sarah.

"We have just one piece of unfinished business," said Sanj. "We'd like to look one more time for Bana. Your town meeting may be the perfect opportunity to continue our search."

“Sanj,” said Sarah, “perhaps if Bana does not wish to be found, you should search no longer.”

“No, it’s alright,” said Aron. “They can go in.”

Sarah gave him a stern look.

They all turned to walk toward the town hall door. Aron touched Sarah’s arm and thought, *I saw Mud up the street. Go stop her.*

Javi, Aron and the visitors entered the hall, and Javi introduced the visitors to Pastor. Sarah peeled off, hurriedly, to find Mud.

Sing-songy Aron began addressing the townsfolk by confirming that he found the guns in an underground chamber in a tunnel while exploring the river area. He decided to crush the guns under the millstone, return the fragments to the chamber and effect a cave-in using explosives. Unfortunately, he somehow altered the river’s path, at least temporarily. He was hoping this deviation was not permanent. Tomorrow, he would inspect the river and the mill to make sure it’s functioning properly.

While Aron was speaking, the visitors looked up and down each row, at each identical face. The townspeople were all wondering why they were being examined. The visitors were satisfied Bana was not present. They graciously thanked Javi and Aron, waved goodbye to the townspeople, calling them a handsome crowd, and left. Javi and Aron followed them out the door and watched as they headed toward Hamlin. Pastor stayed inside and answered questions from the crowd.

Javi and Aron ran in the other direction to find Sarah and Mud hiding behind some trees, watching the visitors leave.

"It's over," said Javi. "Aron was right to let them look one more time—now they know their search is complete. They have no reason to return."

Mud didn't say anything. She had a blank stare, an alarmingly blank stare.

"I'm certain of it," said Javi.

"Please, do not be afraid," said Aron.

But Mud's facial expression did not change. She seemed to be in shock. Then she slowly stepped away from the tree, pulled a gun from her pocket and aimed it towards Sanj's and Clare's backs. Her hands were shaking.

"No!" yelled Sarah.

Out of nowhere, Caveman jumped toward Mud and grabbed the gun. Sarah gasped in disbelief.

"Mud," said Sarah trying to lower her voice. "What are you doing? You can't shoot them!"

Mud broke down in tears and dropped to the ground.

"It doesn't matter what he did to you," said Sarah. "This is another time and another place." Sarah tried to put her arms around Mud, but she was shaking and inconsolable. Sanj and Mayor Clare were far enough away they didn't hear the commotion and had continued on their way.

Aron was similarly in shock to see what Mud tried to do. "Sarah!" he exclaimed. "It indeed *does* matter what that man did to her, a child. No worse sins could've been committed. You should not have said to her that it doesn't matter. Her attempt to shoot him is understandable and forgivable."

Sarah gasped and found it hard to accept his supposition. Her gaze fell to Caveman. "How did you know?"

"Saw Mud. Steal. Gun. Tunnel," she said.

Sarah gingerly took the gun from Caveman and realized the ground tried to reclaim Mud because the gun was in her deep pocket. Sarah held the gun up so Javi could have the same revelation. Javi nodded.

"So you did not know about this," confirmed Sarah looking at Aron.

"No," Aron said, as if annoyed. "And it is none of your business. You are not mayor yet."

"You must've known," she quietly insisted.

"No. I haven't communed with her mind in a long time. Her trauma is so severe that I decided to distance myself to provide her with space and time she needs. When she's ready to talk about something, we talk."

Mud cried out in gasping sobs. Aron squatted down to put a hand on her shoulder. "I am trying to help her as best I can."

Sarah asked Aron, "Was it you alone who found the guns and destroyed them?"

"Yes," Aron said. "It was all me."

This lie was one Aron could keep forever because it came straight from his heart to help Mud.

"Is that the last gun?" asked Sarah.

"Yes," said Aron. Mud nodded.

"Will you destroy it?" asked Sarah of Aron.

"Of course," said Aron. "Caveman, can you please take Mud home?"

Caveman scooped up quivering Mud and left.

Sarah's mind was spinning with shock and anger that Aron thought Mud's attempt to shoot Sanj was understandable and forgivable. *No, that can't be right.* She remembered a verse from the Bible—*Vengeance is mine saith the Lord.* God does not want us to take revenge on anyone.

Then Sarah realized what Aron was doing; keeping Mud and Sanj separated so Mud did not have to worry about meeting him. Yes, of course that was the thing to do. Mud would require a long time to forget her previous life. Aron was going to make sure Mud had that time to forget and heal.

Sarah felt bad she was trying to expect Aron to have Earthling-like or Christian behavior.

Forgive me, Aron," Sarah said. "I was not correct to say what I did. This is not Earth."

Aron hid the gun in his pocket and walked into the town hall to find Pastor still fielding questions about the tunnel. Some people were angry that Aron had kept these secrets.

"We are never going to be a town of united people as long as secrets are kept and distrust festers," said one person.

Aron proceeded to explain that he had found the tunnel but never hiked in it. He said he never heard of anyone traveling that route, until he found out today that Mayor Clare and Sanj from Hamlin tried to traverse it and never saw the arsenal. He did not reveal that Javi, Sarah and Mud knew about the tunnel and decided that was a good secret to keep.

Aron had revealed only to Sarah and Javi that Triol helped him demolish the weapons. That was also a good secret to keep, for Triol's sake.

Aron could live with the disdainful looks he received and mumblings about his character. He would do anything to protect Mud and give her the time she needed for healing and learning to trust, forgive, and love. The reservations about his character would fade over time, he was sure. Aron was the only person who knew Mud's truth—that she had run away from Sanj in Hamlin via the tunnel. Aron found her injured and wandering in the hills behind the mill. He found a home for her, suggested she change her name, and nurtured her toward normal friendships, all to help her recover from her tragic life. Sarah and Javi only suspected Mud was Bana. They didn't really know definitively until she pointed the gun at Sanj. That was another good secret they would all keep.

Aron, then, was the one person who could safely keep Mud's struggles confidential. Even mind readers from Lanet and other planets could not penetrate Aron's thoughts and memories. He had practice resisting. They could read Mud's mind, if they wanted to. But Aron suggested they not. During her introduction to the town, he explained that she was fragile and needed privacy. The beautiful mind reading and empathic people respected his wishes.

Pastor continued mediating questions and discussions from the crowd until all were satisfied that the truth was told. It was a good enough truth. Some people held reservations about Aron's character, but most knew this distrust would fade over time in their peaceful hearts in their beautiful town on Resta. For some joyous persons, their hearts were already mending.

Chapter 22

Triol

As the town focused on guns and tunnels, other activities, such as sports and recreation, were abandoned, seemingly unimportant compared to the captivating crisis just averted. Since no one visited the golf course to notice that Triol was not there, a couple of days passed before it was realized that ke did not appear at kes favorite dinner restaurant one night. Triol's house door was knocked upon and the golf course huts searched, but there was no sign of Triol. Word spread that ke was missing.

Aron, Sarah, Javi and many others divided the town, golf course, fields, and farms into a grid pattern. They each searched their assigned section. Another group planted themselves in the meditation room, and calmly searched the Unity of Souls for any trace of Triol.

When was the last time anyone saw Triol, the golfer from Algenia, Javi's planet? No one could really remember. But Aron knew: it was after they blasted the armory—two days prior.

Sarah grew increasingly worried over Triol's safety since she knew that the ground could reclaim people. She had seen it happen. Well, at least the ground tried to suction up Mud.

Caveman. She needed to find Caveman. Maybe she could help. She guessed she'd be at the manufacturing plant, so she ran there and was shown downstairs to the technology center to find her sitting at one of 50 or more stations, each with monitors and multiple knobs and buttons. There were big cables laying along the desks, down to the floor and through holes in the walls. It was a spider web of conduit and boxes, a complete contradiction to the natural world above.

Sarah did not know what the purpose of this place was, but did not have time to study it now.

"Caveman, please, may I talk to you? I'm so sorry about the other day. Can you forgive me?"

"Nothing to forgive," she said.

Sarah was so surprised at her response. She felt unworthy to receive it but needed to press on. "We have a big problem. Triol is missing. Is there any way you can find her?"

"Allow negative. Pastor," said Caveman.

"The Pastor doesn't allow it? Then there is a way! Every person is important. We can't just let them disappear."

"Lose job," Caveman said.

"No, you will not lose your job. I promise."

Caveman absorbed all this information, thinking about how life could change if she helped Sarah find Triol. Pastor may be angry and hate Caveman for it. Caveman would not be able to stand the intimidation, the rude comments. What would Boulder and the others think if she was degraded to a lower status? Would she be able to enter the Unity of Souls? Could she become an old soul?

What is she thinking about? thought Sarah. *How long will this take?*

"Scared," said Caveman.

"Oh no," said Sarah in her most comforting voice. "Do not be scared. We will solve every problem this town has and make it safe and secure for every person. That includes you and also Triol."

That was sounding like a flat out lie. How could she promise something like that? She couldn't. She put an arm around Caveman's shoulder. "Am I allowed to do that?"

"Okay," said Caveman.

Caveman pulled up a map on a large monitor. It showed the town with a thousand green dots. Caveman pointed to the largest dot and said, "Person, Frond, Ganead. Eliminate all." She pushed a button and all of the large green dots disappeared.

Sarah gasped when her first thought imagined all those people disappearing. Then she quickly calmed herself and became certain they were just being sorted off the screen.

Caveman pointed to a small green dot and said, "Person, Lanet, Tupit, Mexopin, and Melotin." Upon pushing another button, all of the small dots disappeared. There were many different sized dots Caveman could identify and continued eliminating populations from view. Seven dots remained.

"Algenia, Velam, Euronius," Caveman said. Sarah scanned these dots for location. One was on Main Street. A few were isolated off to the side, presumably in their houses. Two dots were on a hiking trail. And one was beyond the golf course, on a small road, almost at the edge of the monitor.

"There! Could that be Triol?"

"Yes," said Caveman.

"Thank you, thank you!" cried Sarah. She gave Caveman a big hug and dashed up the stairs, through the plant and down to Main Street, but she didn't see Javi or Aron. She headed for the golf course alone.

Sarah raced through the golf course, found the road and kept running. After three hours, no Triol. After three additional hours, still no Triol and no other people or town. *How big is this planet?* She felt deflated and resolved it was time to return home. *Home*, she thought. *I like the sound of that.* She turned and headed back, thinking only of Triol, a respected miner from Algenia, Javi's planet, where envy was considered a good character trait. If Triol is so respected, what would ke covet? Who would a miner envy? Persons who lived above ground in the sun?

Persons who were not under the threat of cave-ins? Maybe ke was not of good character and was not envious of anything. Maybe ke just wanted to be free and play golf.

Finally deciding it was too dark to proceed, Sarah laid down on the side of the road. She looked at her hand and saw the purplish haze. She blinked, and it was gone.

Sarah fell asleep and dreamed someone was gently saying her name.

"Sarah. Sarah."

"Triol!" exclaimed Sarah. "I have been looking for you. What are you doing out here?"

Ke sat beside her. "I can't keep a secret," ke said.

"What?"

"I decided to leave because I can't keep a secret the way Aron can," ke said.

"It's okay if you come back and someone discovers you helped Aron destroy the armory," said Sarah. "Probably no one will read your mind. But, truth is important, too."

"That is why you will make a great mayor. But truth is second to Aron's burden. He is shouldering the most selfless act I have ever seen. Leaving Sedona is the best way I can help him," said Triol.

"I see what you mean," said Sarah. "Aron is using his faith, body, and soul to help Mud. I have only made it more difficult for them. You know, Triol, I tend to think that everyone would accept each other if we all acted like Earthlings. But that's a huge mistake. There are many cultures and religions through which God blesses us. I am seeing that I have a weak faith. I need to spend my time praying and loving. Triol, I can't keep secrets either. Anyone from Lanet can read my mind. I have no way of stopping that. I am beginning to understand that people from Lanet are the most discriminated against because others are afraid of their mindreading abilities. But they are respectful people. That's the most important thing." Then very deliberately, she said, "But we can still figure out how to live together."

She stopped short and realized this was the most important thing she had ever said. *With all our differences we can still learn how to live together*. At that moment, Sarah knew what her passion was and what contribution she could make to this world. She thought for a moment. "Why did you wake me up?"

"To let you know I'm safe. Thank you for finding me," ke said.

"Will you come home?" she asked.

Ke nodded, and they got up and jogged home.

In the town hall, people were asking Pastor about GMDisruption and how it could have swallowed Mud. Pastor explained that the force can be adjusted to include, or exclude, any sized object. The intensity can also be adjusted to allow for various metal content. There must've been some magnetic rock under Mud's feet to cause the nerve-racking incident. A reasonable explanation.

Sarah and Triol walked in, and ke received a lot of hugs and pats on the back to say *Welcome home!* Aron grabbed Triol from the back in a great big hug and said, "I am so relieved! You scared me to sadness! I would never forgive myself if you didn't come home." Aron mouthed *Thank you* to Sarah. Several people patted Sarah on the back and expressed their gratitude and said they were impressed she found their dear friend Triol. No one asked why Triol ran away. *How gracious*, thought Sarah, *not to question ke, just to be grateful ke returned.*

Sarah wondered about Pastor's imagined story of possible magnetic rocks under Mud's feet as a plausible explanation for sinking. Did the Pastor truly believe that or was she just trying to sell it to placate the town? Well, either way, Sarah thought it was a pretty good story. Perhaps she knew the truth about the gun in Mud's pocket. Who knows? Anyway, when people join in the Unity of Souls, won't the truth be revealed? She doubted that even Aron could keep secrets forever.

Pastor addressed the crowd. "Aron nominated Sarah to be mayor because he saw in her what I see—a person without prejudice who loves meeting every person and learning about them. If everyone were like that, Sedona would be a more loving community."

Actually, Aron's opinion of Sarah ran hot then cold. She was bossy, then soft-spoken; overbearing, then gentle. But she did have a way of getting everyone to speak calmly at town meetings. That was good. She would be able to address big issues. Hopefully, there would be no more big issues. He wished she did not feel obliged to befriend everyone. Was that love or meddling? Probably love, because it was a gift God gave her. Just as God gave Mud to Aron to care for, He gave the town to Sarah to care for.

Then Aron said, “With no objections, I award Sarah the position of mayor of our beautiful Sedona.”

“I object,” said a voice.

Everyone looked toward the back of the hall and saw strange, angry man with a red aura. Sarah had to think for a moment. *I know. He accused me of spying on people.*

“I’m not leaving Sedona and I’m not getting dragged into the Unity of Souls. You can’t force me because I won’t do it.”

“Of *course* you won’t be forced to do anything,” said Sarah.

“I can’t trust you,” he said.

Javi moved over to put an arm around his shoulder and said, “But I know her. She keeps her promises.”

Strange, angry man scurried out of town hall.

Pastor looked at Sarah and said, “He’s all yours.”

"Now," said Aron, "if there are no more objections, I award Sarah the position of mayor."

Everyone clapped and cheered. Sarah bowed to them, wondering if that was the correct expression of gratitude. Javi cheered the loudest. Ke was very happy and not envious at all.

CHAPTER 23

HOW TO BE MAYOR

Sarah, Javi, and Pastor sat in a café the next morning, sampling various teas and pastries. Sarah congratulated Pastor on excellent moderator skills.

"I was quite nervous about it," said Pastor. "I normally converse one-to-one."

Sarah wanted to initiate a conversation about how Pastor could say Voder had a pea-sized soul, but she was not bold enough. Voder returned to Hamlin, so what did it matter now.

"Where are you from?" asked Javi.

"I'm from Far, a female," she said.

Sarah was stunned by that simple sentence. "Your planet's name was Far?" she asked.

"Yes," said Pastor.

"I am such a moron! I asked Caveman where she was from and she said 'Far', but I assumed she meant her planet was far away."

"No, that's the planet's name," said Pastor.

"I am an imbecile, a total moron," said Sarah, shaking her head.

"I didn't realize," said Pastor. "You hide it very well."

"No, I'm not *actually* a moron, but I sure feel like it sometimes."

Javi sat there smiling at the conversation.

"Pastor," asked Sarah, "why is your speech so much more eloquent than the others?"

"I learned to construct sentences properly when I came to Resta. I wanted to be a pastor and needed a better command of language to do my job effectively. Don't misunderstand Caveman, Boulder, and the others: they are highly intelligent."

"I see that," said Sarah. "I have been nothing but rude and offensive toward them. What a poor representation of an Earthling I am!" Sarah hid her face in her hands.

"You have also been delightful and funny and…" started Javi, but Sarah interrupted ke.

"Don't say nice things about me, please. I want to be miserable for a few minutes."

Pastor gave Javi a questioning look, but ke just shrugged.

Sarah's head popped up in surprise. "Pastor," she said. "How large an object can the ground absorb?"

"Probably one cubic foot. Why?"

"Aron went through a lot of trouble for nothing. He didn't need to demolish the guns. The crates could've been broken and the guns absorbed whole."

"Interesting," said Javi. "No one knows how this environment is controlled, so no one knows what options are available. We concentrate on the spiritual and ignore the physical."

"That is for the benefit of all," said Pastor. "The fewer people that know about the technology center the better. It is for their protection. If someone would've told me about the armory, I would've taken care of it."

"But no one knows that," said Sarah. "No one, except those from Far, know there is a way to take care of it. No one knows that you are the person who knows of the secret technology center."

"It's not that it's truly secret," said Pastor, "but why bother about it when there are other more important things to be concerned with. In addition, we don't want many persons trained to be operators to prevent someone from modifying Sedona to resemble their hometown on their home planet."

Javi and Sarah looked at each other acknowledging that was a good point.

"Well, then," said Sarah, "we need to encourage people to focus on spiritual matters and develop deeper friendships."

"That has always been our goal," said Javi, "and we need to encourage it even more, since our town has been shaken awake."

Sarah told Pastor Voder's story and about him returning to Hamlin. "We were wrong about him and the armory," said Sarah. "He knew nothing about it. And he didn't care about it."

"I feel badly thinking ill of someone," said Javi. "It's as if I can kill him in my heart."

"Perhaps we all sensed his inner turmoil, his guilt concerning his daughter," said Pastor.

"I'm glad we found the armory," said Javi. "It shows us that there were previous inhabitants that became militaristic and now they're gone. I wonder why they felt the need to defend their town or attack another. They must've made some mistakes. We are at least the second civilization to come to this area."

"I think you're right," said Sarah. "And that means this cannot be Heaven. Now, I really think we should address the issue of missing persons."

Pastor interrupted and said, “Mayor Sarah, why don’t I give you a short tour of the Tech Center. You’ll need to know the basics.”

Pastor took Sarah, but not Javi, to the Center to explain briefly how it worked. Sarah learned that the ground was underlain with a network of sys-paddles and sensors linked to a mechatronics system. Servomechanisms controlled soil distribution, river flow, and GMDisruption, to name a few. There were also sensors for detecting various animals, humans, and chemicals. Long rows of monitors controlled all these systems. Boulder was at the hydrology station with Bull looking over kes shoulder. Ten or 12 other workers sat at their stations. Pastor did not acknowledge them.

“The current operators are from Far, led by Caveman and Boulder,” Pastor said. “They do a good job and will serve you well. Feel free to ask me any questions you may have. I am always willing to share my expertise.”

“I wonder why you don’t allow the Tech Center to help me find missing persons,” said Sarah.

“To allow them to leave and not feel like they will be dragged back,” replied Pastor.

Walking to the church, Pastor said, “One of the most important responsibilities for a mayor is to make newcomers feel welcome and help them acclimate to their house and surroundings. This is where I feel you will excel.”

“Thank you,” said Sarah.

“You must also ensure the town has a pastor. I assure you, I am happily residing here and have no plans to leave. Now, you can run the town any way you like, but I want to tell you my philosophy: souls are more important than bodies, and the development of old souls is more important than the community.”

“Huh?” said Sarah.

“Souls are more important than bodies, and the development of old souls is more important than the community. The purpose of Resta is to provide a place for people to develop awareness of their souls and the maturation of their souls. Everyone’s goal is to develop an old soul and to join in the Unity of Souls.”

Sarah understood all of this, but felt something was missing. Something didn’t sit right. She thanked Pastor and left, anxious to talk to Javi.

CHAPTER 24

WATERFALL CHURCH

Sarah conducted a fact-finding tour. She asked whomever she passed how many people in Sedona came from their home planet. She found that one came from Urag, three from Tupit, one from Mexopin, two from Melotin, at least two were from Earth, one was from Lanet (Aron) and two were from Algenia (Javi and Triol). There were one or two from a bunch of other planets. But she saw 20 that came from Far and they were all Tech Center workers. That was the only anomaly.

Even though Pastor gave an explanation why lost persons were not found, Sarah felt Far operators were afraid of Pastor. When Sarah asked Caveman to help her locate Triol, Caveman had thought about it a long time before responding. Sarah was certain no one should be scared in Sedona, and everyone should be equal.

Why would Pastor engage in societal problems from Far when the slate was wiped clean on Resta? Because she was not really a pastor and had ulterior motives? Because Caveman and the others succumbed so easily? Or because she'd stop at nothing to elevate herself in the Unity of Souls or in Sedona? *She is not a spiritual person despite her pretty purple aura,* thought Sarah.

Sarah ran back to the Tech Center to talk with Caveman, but she, Caveman, was not there. Sarah jumped up in the highest tree and looked all around, spotting her behind the park at the base of the mountain. There was a waterfall curtain Caveman and several others were going through. Sarah sprinted then sidled up to the cave and peered in surreptitiously just far enough to see Caveman and six others from Far on their knees, hands clasped in prayer. Their rainbow of auras was rich and stunning. One of them with a blue aura looked familiar, but Sarah couldn't remember.

Sarah realized what they were doing; worshiping God in their own little church. Sarah turned her head and listened to prayerful mumblings and a soft hymn. How revealing. They didn't want to go to Pastor's church so they dedicated space for their own. Sarah waited for the hymn to end then walked through. Caveman and a couple others jumped in surprise.

"I'm sorry," said Sarah. "Don't be afraid. I just want to talk…to know what I can do to make life better for you here."

Caveman said, "Nothing. Pastor mean. Always be. Only job Tech. Order us control Sedona. Geography, ecology."

"Let me see if I understand," "said Sarah,". "You think nothing can be done to make life better for you because Pastor is always mean to you and will always be mean to you. The only job she will allow you to have is Tech Center worker so she can command how you maintain geography and ecology. But why do you have to do what she says? Don't you already have training and expertise in this?"

"Must," said Caveman.

"No, you don't," said Sarah. "The society on Far is not the society on Resta."

Their eyes expressed surprise, as if they never knew that.

"I am the mayor now. Pastor acts like she *wants* me to be mayor."

"Control you. Think you weak," said Caveman.

"Oh, dear," said Sarah sadly. "She thinks I don't care about each person. She thinks I am weak and will blindly follow her or be manipulated by her." Sarah squared her shoulders. "I will show her. In the meantime, don't do everything she says."

As Sarah left, they glanced around, smiled, and someone said, "Verbose."

CHAPTER 25

MISSING PERSONS

Later, Sarah found Javi was seated alone in the same restaurant where they had met earlier.

"Can I join you?" Sarah asked. "I really need to talk to you."

She spewed out about the trip to the Tech Center and how every controller was from Far. Generally only one or two persons outside the Tech Center were from each planet, but at least 20 represented Far. She lowered her voice and told ke about her conversation with Pastor. "I just don't feel right about her views, that souls are more important than bodies, and the development of old souls is more important than the community. *All* these things are important. People with all their souls together make up our community. I am concerned that she excludes people who are new in their spiritual journeys. When I first came here, I had never meditated or learned how to pray properly. But she didn't exclude me."

"Who do you think she excluded?" asked Javi.

A person walked by the table, so Sarah waited before answering.

"I heard of two people who couldn't find anyone else from their planet. What if Pastor got rid of them? Told them they weren't welcome here?"

"I thought those people left of their own accord," said Javi.

"That's what she would tell you. I don't think she values every person. I think every person who sets foot in Sedona should be welcomed and befriended…the way you did to me."

"Did Pastor ever try to usher you out?" asked Javi.

"No, not at all. Maybe she believed my soul would mature. The only time I ever disappeared unexpectedly was when I transcended into the tunnel."

Sarah stopped short. She never understood why or how she transcended into the tunnel. "Is it possible Pastor *made* me transcend?"

"I don't know," said Javi thoughtfully. "I've never had anyone make me transcend. So she put you in the tunnel hoping you'd wander off and live somewhere else?"

"Sure, why not," said Sarah sarcastically. "Maybe she didn't realize you would come find me. You really did save me. Or…maybe she didn't know the tunnel was there."

"She sent you into the ground?" Javi shook his head. "That's terrible. That implies severe mal intent on her part."

Sarah's mind was racing, her eyes darting back and forth. "I wonder if Aron can read her mind," she said. "Can we find out what she's thinking in the Unity of Souls?"

"I think Aron could read her mind. But the Unity of Souls is not about minds."

"How many people have disappeared from Sedona?" she asked.

"I've known of six since I've been here," said Javi. "One probably ran away due to a disagreement with Pastor. Ke wanted to run the technology center, but Pastor said no. Two are probably the ones you mentioned who wanted to find people from their planet. And there were two who left but I know nothing about it. And then there's Paul."

"Was Pastor functioning as our mayor then?" asked Sarah.

"Yes."

"And she ran the Tech Center and still does."

"Presumably."

"You don't suppose she sank *them* into the ground, GMDisruption," suggested Sarah.

"Buried alive?" asked Javi. "That cannot be. Your imagination, Sarah!"

"It's a horrifying thought. Please forgive me," said Sarah. "In the tech center, items can be raised and lowered in the ground—you know that capture game. Maybe it can be done with people."

"Sarah, you are the only person I have ever known who cares so deeply about lost individuals." Javi put a reassuring arm around her. "I don't understand it. I always assume that each person is safe and has the freedom to pursue whatever they desire without reporting to anyone."

"That's the way it should be," said Sarah. "But I'm from Earth. If anyone disappears, we assume there has been a violent crime. We all join together and hunt for them and pray we can find them alive."

Javi was speechless, clearly unknowing of such criminal activity on Earth. "I agree now that we need to find where these people are and if Pastor is involved."

Aron came quickly into the restaurant and spotted Sarah and Javi. He seemed agitated. "I can't find Dora!" he said.

"What?" said Sarah.

"Help me look!"

Aron went back to the mill, and Javi started checking all the restaurants. Sarah ran right to the Tech Center.

"Caveman! Dora is missing! Can you find her?"

"Afraid," said Caveman, glancing toward some of the other operators.

"Oh, no. Don't be afraid. Your help in finding missing persons is commendable. We couldn't've found Triol without you. We need you and your valuable expertise."

Caveman stared off into space for a bit, then did her magic, and Sarah ran to find Aron and Javi.

"To Hamlin! Dora's on her way to Hamlin," Sarah yelled to Aron.

Sarah and Aron took off like horses out of the gate. After an hour of high speed pursuit, they saw Dora, shuffling slowly on the road quite a distance away, her head drooping.

"Dora!" Aron yelled.

Dora turned and was very surprised to see her friends. She smiled a big smile and ran back to them.

"We have been looking for you everywhere!" said Aron. "Are you alright, my dear friend?"

"I'm sorry I did not say goodbye," said Dora. "I feel so lost, I had to leave Sedona immediately."

"Why? Why?" asked Aron." Do you need other friends or a new job? How about a vacation? Are you angry with me? Did I say something wrong?"

"No, no, everything was fine. It's just…it's just that Pastor said I will never grow spiritually in Sedona. That I will never be able to join the Unity of Souls."

"Ignore that. It means nothing!" exclaimed Aron. "I don't think this has to do with the Unity of Souls. I think it has to do with the commune of minds, that is, all of us with this gift from Lanet, Narth and other planets. Pastor considers this ability powerful, threatening and making it more difficult to develop an old soul. But that's not true."

"Why does she say that?" asked Dora. "I don't threaten anyone. We are all nice people. And our goals are still the same as everyone else."

"I know," said Aron. "But I'm suggesting that Pastor may not see it that way. Pastor…" He hesitated, dreading revealing this tidbit of information. So he spit it out quickly. "Pastor wants all residents, 100% of us, to develop an old soul. Then Sedona may be considered as a site for a Temple for God, a Holy Temple for God."

Dora drew a deep breath and sighed in amazement. "Alright let's go back."

They arrived in Sedona and filled in Javi on their discussions.

"Pastor asked Dora to leave," explained Aron, "because she needs 100% of souls here to be old souls. Then Sedona can be considered as a site for a temple for God. Pastor believes Dora will not be able to strengthen her faith, not her faith."

"A Temple for God!" said Javi. "Why do you think this, Aron?"

"Because I read her mind. Once. I am certain she wants no one discovering her plan by reading her mind or by any other means."

"Why did you read her mind in the first place?" asked Sarah.

"I desired to determine why a person had an argument with Pastor and then left. And I discovered Pastor felt this person had no chance of building a faith-filled life and would never develop an old soul. Pastor, though, told the person that a job at the manufacturing plant or technology center was not possible, and ke should leave. Relocate to another town. Maybe this is the reason Paul disappeared— Pastor asked him to leave, too."

"That explains a lot," said Sarah. "I also want Sedona and all of Resta to be a wonderful place to worship God. But I don't think we should sacrifice people for it…for a temple. Dora, I'm so glad you came back. If you desire to grow spiritually, you will have all the time you need, and all the support you need, from whomever you want. But this is not because I just proclaimed it or you need my permission, but because it is what you want."

Dora sighed. "I really did not want to go to Hamlin and see Voder again. I probably would have passed through and kept going. I'm so relieved!"

"Group hug!" exclaimed Sarah, reaching her arms out.

“I love it!’ said Aron, embracing them all.

What’s this? thought Javi, but ke joined in.

CHAPTER 26

DISCOVERY

Later, Sarah ran the mountain a mile from Sedona just to unwind and get a new view. It was taller than the mountain next to Sedona and had tons of snow on it. Numerous waterfalls crashed onto a rugged landscape below filling a lake and providing the headwaters for a river running vaguely in the direction of Hamlin. Elk, camels and unknown creatures with short legs and thick necks grazed on the shores. Across the lake and valley, additional mountains rose to meet the sky. Sarah laid back, hands behind her head with fingers locked, and she wondered why she could never see the sun. *Is the Brightness of the Beyond a mechanical device? A dome. Like a giant light bulb with a dimmer switch? Yes,* she continued thinking that could be true.

She felt a small tremor and knew what it was – some GMDisruption going on. Somewhere. Perhaps Aron, Triol and Boulder were re-creating the river's flow at Mill Creek. She looked across the valley to see mountains with orange peaks, and purple and gray striations symmetrically formed. *How is that possible?* In a place where there are rivers and blustery winds, there should be some worn peaks just as on the mountain by Sedona and the mountain on which she was sitting.

The tremors increased gradually and she witnessed the distant striated mountains' heights increase. She sat up more erect to get a better view. The level of the lake decreased. The striated mountains added even layers of rock at their bases, either gray or charcoal in color, thus increasing the height of the mountain peaks.

Is there a tech center underground? Sarah wondered. *This can't be natural geology. There's no way the mountain peaks would rise so evenly. Would a hunt for a tech center be in order?*

Oh God, please tell me what's happening, she muttered. *Tell me what you want me to do*. Clearly an insincere prayer.

A presence slowly developed in Sarah's head. It said, "Bring Voder back. Bring Zac back. They are ready to rule."

Sarah jumped up yelling, "What? What?" Who are you?" Leave me!"

"We are the gods of prejudice, indignation, mistrust, and tension," they said in her head.

"No, there's no such thing! You are demons! Go away!" she demanded. She began feeling woozy, the way ether makes one feel when having wisdom teeth extracted. She felt her mind traveling thru a rotating cone of alternating black and white sections. The demons tried to hold on, but Sarah fought them off. She was unsure what happened to them. As her dizziness dissipated, she could hold her head up and look around. Did they disperse into a waterfall, or a bird or a camel or a pig that just ran into the lake? The earth tremors ended, and the lake and mountains looked stable. Sarah laid her head in her hands and started to cry. In this world, nothing was ever like Earth. *Why couldn't it be like Earth?* On Earth there were no voices in her head, no requirement to discern right from wrong. The rule of law applied to everyone. Philosophers philosophized, and the justice system justified. She had never read a book or heard a voice telling her there were gods of prejudice, indignation, mistrust, and tension. Were these the things that affected people in town? She had already seen prejudice in herself and mistrust in Aron and in the angry man from the forest. Of course Mud was filled with tension and anxiety. Was she admitting that she would not be able to help people with their difficulties? What was Pastor doing all this time if not helping people with their problems?

Sarah wondered if she would actually be a bad mayor because she didn't really treat people with all the dignity and respect they deserve. People from all different planets needed to be treated with the respect they would get on their home planets, and that included Voder and Zak from Frond. And Aron, so loving, so passionate, needed to be encouraged to continue his mission to protect Mud: even if he was irritated by Voder's egregious behavior. She never thought about demons or evil. She had never interacted with people who believed in many gods. Were there truly many gods? She discarded that notion as ridiculous. Wait a minute. There must be religions on Earth that believed in many gods. Yes, she remembered that! Hades is the god evil. Asmodeus, in Jewish legend, is the king of demons. Then there were millions of Hindu gods. In Christianity there was one God in three parts.

And Judaism and Islam each had one God. But in Hinduism there were millions of gods. Still, on Resta, even though she believed in the Trinity, she would be more aware of any possibility.

Sarah pulled herself together, wiped off her tear-stained face and decided to investigate the perfect triangular mountains by hopping across the valley and doing a little exploring. It's amazing how much courage she gained since developing the transcendence ability. She jumped down the mountain and took a giant leap across the lake but misjudged its width. She landed in four feet of water and laboriously made another leap to the shore. Going home to change seemed preferable to exploring, so she squeezed out as much water as she could from the bottom of her robe and boing-ed until she found the road and went home.

She changed robes and knocked on Javi's door to tell him of her run. Ke thought it would be very interesting to know of other possible tech centers. They set out for the perfect triangular mountains. The ground and smooth rocky cliffs of the mountains were quiet. Sarah and Javi hopped slowly up the sides, all the way to the top.

They heard faint whistling winds like birds in the distance. The sounds were pure and seemed to make melodies.

Sarah said, "Javi, I think those are hymns to God."

"Then are those angels singing?" asked Javi.

"Could be," said Sarah. "I hope so. It's so beautiful."

They sat on the wet snow, their auras entwined, and they listened to them.

Gradually, more rumbling could be felt and then heard.

"Where?" said Sarah.

"Sedona," said Javi. "No, the mountain by the town."

They jumped up and boing-ed toward Sedona's tech center. The residents saw them hopping home and yelled to them, "Don't worry. Bull has devised a solution for the mountain decline."

Sarah and Javi didn't know what they were talking about, so continued to the tech center. Bull was heading up discussions. All the stations were manned and whirring as Sarah had never heard before.

Bull saw Sarah and Javi and said, "I don't have time to talk now." They walked back up to the surface and joined the crowd. Triol saw them and said, "The mountain's base is being smoothed out, and three waterfalls are being combined and routed toward our river. Not only will that restore power to the mill, but allow more even ground on which to build houses."

CHAPTER 27

CONFRONT PASTOR

Sarah decided she wanted to talk with Pastor, not knowing what to say but hoping the right words would come out. Javi said Pastor was not at the church so they hopped back to the Tech Center. They walked slowly down the stairs, crept to the basement door and remained in the stairwell to eavesdrop.

They heard Pastor say, "I'm controller, not Sarah. Understand?"

"Yes," the workers from Far said.

"Not search," she commanded.

"Not search," a few mumbled.

"Not search," she repeated loudly. "Caveman?"

"Not search," said Caveman.

"No!" said Sarah, bursting through the door. "You are not the controller! Neither am I. And Caveman is not your servant. None of these people are your servants!"

Pastor squinted her eyes and slowly shook her head. "Of course they are not my servants. They work for the good of our community."

Sarah said, "They can listen to anyone's opinion and base their actions on that and their own knowledge and expertise. And as far as searching is concerned: their searching ability was instrumental in locating Triol and Dora. And how could you tell Dora to leave? You have no right."

"It was good advice for her, for her growth," said Pastor, fairly menacingly.

"No, it was manipulation by you, and it was wrong. This town is very important, and each person in it is very important and needs to be loved. I know it. I feel it," said Sarah, running out of breath. "There are others who left. Did you ask them to leave, too?" she asked, not expecting a response. Slowly, she said, "I think *you* are the one who needs to leave."

Pastor and all the workers gasped. That statement took their breath away. No one had ever talked like that to Pastor before.

Sarah said, "Now you see what a horrible statement that is—how it crushes the heart. I have no right to ask you to leave. What I'm asking from you is a little humility."

Pastor was motionless. Then she explained in her most comforting voice. "This is about you and me collaborating for the good of Sedona, for the good of each person. Together you and I can make Sedona blossom into a great haven of old souls. We need to accomplish this together."

Sarah shook her head and said, "You are insincere and twisted."

Another gasp from the workers.

"Grmph, Grmph." A low-frequency throat clearing came from the back of the long room. Sarah watched a person walk slowly forward like a bride down the aisle in a sparkling white robe and many silver sashes around the waist and neckline. It was Bull with a delicate red aura. He said, "Sarah, we provide best operation of Center, for comfort of all. Operators under management of Pastor have greatest degree of excellence in this facility. Sedona has been and will remain perfect town. We believe Pastor provides best leadership above, as well. But I not aware of Triol or others."

Sarah said, "You have given me no reason to think others can't be trained to work here. At least one other had the desire, but Pastor refused."

"Don't you want to know who I am," he asked.

"Bull," said Sarah."

"And my title," said Bull.

"Sure."

"I, Bull, first one here. Prince of Ferista race, Sovereign god of Far people. Sent here to prepare way for you." He tipped his chin down and eyes glaring up at Sarah just to exhibit a sense of importance as to what he has accomplished. But Sarah didn't care.

"I don't think you are the Sovereign god of Far people when you and they are on Resta. God Almighty is the Lord of everyone, even you. You need to get to know Him. What do you mean by 'you are the first one here'?"

During Sarah's question, Bull's aura flared and erupted into uncontrollable waves. Sarah took a step back not knowing how swollen he would become. The anger on his face mimicked a snorting bull and would've been comical had he not thought of himself as a lower case god.

Sarah repeated her question. "What do you mean by 'you are the first one here'?"

"I am from Far's past civilization when war was rampant. Those civilizations were eliminated. I asked for other Farians to come help synthesize environment so you can concentrate on spiritual development," said Bull.

"Why didn't you destroy the weapons?" asked Sarah.

"A cache in case they were needed. But Aron seemed determined to eliminate them, so I allowed it."

It took Sarah a moment to absorb all this information. Then she asked, "Is Pastor truly a pastor?" Another shot to the heart. Sarah was surprised how easily she could hurt people with her words. She wondered if it was necessary for her to do this. "And why are you so secretive?"

"Yes," said Bull, "she is a pastor and I sent her to surface, assume do good job. I also go to surface but didn't know people disappear and didn't know we appeared secretive. I will become more visible as leader of Sedona and assume prominent role in business and religion."

Sarah said, "I have been assigned the role of mayor. Did you know?"

Bull remained silent.

Sarah continued, "Yes, people should get to know you, as the Sovereign god of the Far people, but not as a Dictator. And Pastor needs to stop telling people to leave. Also, others from Far don't know who to listen to," said Sarah. "Do you know about praying behind the waterfall?"

"Yes," said Bull. "So?"

"The purpose of that place is to be able to worship somewhere that doesn't have a caste system," said Sarah.

All the Far workers groaned and put their heads down on their arms or keyboards. They were rather embarrassed to have Sarah explain their criticism of Far society to their Prince.

"Explain?" asked Bull, glaring at Pastor.

"All deepen spiritually," said Pastor, "because of me."

"Not all," said Bull. "All afraid. But why afraid?" he asked the others.

"Pastor said not tell," Caveman explained with her head down.

Bull looked at Sarah and nodded.

Sarah said, "Those from Far would like to try other jobs and have Far's social injustices removed. Bull, Resta is a world where, I believe, individuals can be truly free. This might be quite different to you, but I encourage you to communicate with people from other societies, and you will find freedom is precious and priceless."

Bull, unexpectedly, slammed his fist down on a console and yelled, "There is no such thing as freedom! Everyone must belong to something. Humans are like herding creatures—they must have others to survive and a head of the herd. I will always be Far's Prince and god. Nothing you can say will change this."

"Well, how about—the Lord wants us to mature in the Unity of Souls and develop old souls. We do not prance around out here as fake deity," said Sarah.

Bull spun around to see Pastor and growled, "How could you let this happen?" Bull reached for Pastor but Sarah reached her first and transcended Pastor and herself to the surface."

“Why would you care about my position here,” she asked Sarah.

“Because we are all living together,” Sarah replied. “Wait for me here.”

Sarah ran down the stairs to the Tech Center. Pastor disobediently followed her.

“That was a good trick,” said Bull to Sarah. “Don’t ever do that again.” His eyes could’ve shot out little lightning bolts.

“I won’t have to if people can trust you to behave in a civil manner. That’s where the word civilization comes from. It’s great you created this beautiful town as a home for us. Thank you. But now you and Pastor can move on to your next activity of community or philosophy. Write a book on the past civilizations. We can all learn from history.”

“I don’t know, but I will try,” he said, barely getting out the word ‘try’.

“Would anyone like a new job?” she asked the workers gently.

A short silence followed.

“Now?” said one of the Farians.

“Yes. Why not now?”

Ke stood up and looked at Caveman, who also stood up.

They walked to the stairs and looked back. Gradually, others stood and headed for the stairs.

"Please, some stay until new operators trained," said Bull. Bull was glad to see that Pastor returned.

Three operators agreed to stay. They, Pastor, and Bull watched the rest climb the stairs. Sarah turned to say thank you to Bull.

Bull looked at Pastor and said, "Sit."

Sarah, her purplish haze and those that wanted new jobs went outside where she gave them each a gentle hug and told them to think about what they wanted to do.

"I want golf," said Caveman.

"I want…I want…I want…" they all proclaimed.

"I want be pastor," said Boulder.

It was clear they didn't need any time to consider the matter—they already knew.

Javi had not said a word during this entire event. Ke just admired Sarah for maintaining control of this difficult situation.

CHAPTER 28

TOWN MEETING

Sarah the Mayor held a town meeting the very next day. She assured everyone that no government and no laws would still be the rule. Sarah made each person understand that she wanted every person to stay in Sedona if that was their desire, regardless of soul status. She knew now that Resta was not Heaven, but maybe the Promised Land, a beautiful place where joy and spirits grow. It was disappointing to know that below the ground were nuts and bolts, a subterranean network of metal and equipment and motors. Still, on top, souls would develop into the purely loving souls God wanted them to be. Sarah didn't know if Resta was being considered as a temple for God, but Pastor believed it, and she was welcome to her beliefs.

Sarah confirmed the rumors that Boulder would become the new Pastor. He was learning sentence structure from…that's right…the old Pastor. This would take time. Just then, Sarah remembered the first time she saw Boulder, the blue aura person sitting on the grass who had other people sit facing him, and they'd 'talk' for a while and swirl auras. The "psychiatrist" she had called him! What a blessing to have him as the new Pastor. He'd be able to enrich everyone's aura and soulful experiences.

Vacancies opened up in the Tech Center. If someone was interested in a job there, they should go see Bull, the one from Far with the deep voice.

Sarah asked the crowd if anyone knew about Paul, where he possibly could've gone. No one knew. "What planet was he from?" asked Sarah.

"Melotin, I believe," said someone.

"No, I thought it was Earth," someone else said.

"Did he ever join you in the Unity of Souls?" asked Sarah.

"No, never. He never went to church or meditated," said someone.

"All he liked to do was golf," said Triol.

Sarah wondered if Pastor told Paul he would never develop an old soul.

Mayor Sarah asked and then begged people to let her know if they were traveling somewhere. "Please, please, please let me know! If someone tells me you are missing I will spend my time trying to find you. *All* of my time. You are terribly important to the spirituality of Sedona, and I want you safe."

Sarah announced town meetings would be held every other afternoon to smooth over the rough edges of personalities. All people must learn to live together, work on patience, practice socializing, grow, and really talk with people from other planets. She herself promised to control her sarcasm.

Sarah had a true home now. She had the job right for her and truly loved her friends. No one was perfect, except maybe Javi. But everyone in the town belonged together. She said she believed all of soul, body, and community were important.

In the Tall Grass Restaurant, Sarah and Javi dined on oat dishes.

"Why did you name Caveman *Caveman*?" she asked.

"Those people didn't really have names, just identifying marks. So I was providing some clarity for the group to function in society."

"Do you know what a caveman is?" asked Sarah.

"It's obvious. There are many caves on Far, so I thought the name was appropriate."

"Did you name the others?" asked Sarah.

"A couple. Boulder, Sleepy."

Ah, the Seven Dwarfs, thought Sarah.

"I must compliment you on your managing skills, your ability to negotiate and nurture persons for the good of all and understanding of all. I see now that's the most difficult aspect of people from all different planets and cultures living together—that they understand each other, both in communication and in ethics."

"Thank you, Javi. I have learned that people grow in spirituality over time, but at different rates. It cannot be rushed."

"I wonder if Pastor ever asked Mud to leave Sedona," said Javi.

"I doubt it," said Sarah. "Aron would never allow it. Resta truly is an amazing place. I can hardly believe I can transcend now. I feel my body and soul work together to accomplish this. It must be by God's grace." Sarah put out her hand and marveled at her purple aura. Javi's blue aura extended out from his body and intermingled with Sarah's, coaxing it outwards. The two swirled together into a bright blur. The depth of their love was unbelievable, beyond anything either of them had ever experienced. They both felt unimaginably strong. Sarah never wanted to let go, she wanted to stay in this entranced state forever. *Now, this is Heaven,* she thought. She felt Javi engulf her and promise an eternity of love.

However, they slowly separated, their auras unwinding, regaining their separate feelings.

"I felt so strong with you," said Sarah.

"And I with you," said Javi. "Let's go visit your tiger. Do you want to walk or run?"

"I enjoy walking very much," said Sarah.

"You have never said that before. You are pulling my leg," said Javi.

"No, I'm not! I do enjoy walking very much. I just never knew it until I met you."

They chuckled as they began their walk on that glorious day.

Chapter 29

Epilogue

Paul sat miles and miles away at the top of a jagged mountain, looking down over the beautiful planet he loved. He was not missing. It was time to move on. This assignment, Sedona, was a difficult one. Paul taught Sarah to transcend by making her transcend into the tunnel: she was learning God blessed her with this gift. The civilization that stocked the arsenal was long gone. Aron and Triol destroying the armory was quite an accomplishment. Paul did not require that the arsenal be destroyed, but Aron was committed, so Paul let it happen. Mud, aka Bana, was where she belonged, thanks to Aron; General Voder returned to Hamlin to start a familial relationship; the Far people were beginning a more pleasant, interesting life; a new Pastor was beginning his ministry; and Mayor Sarah was happy welcoming each new person and encouraging good communication and trust amongst all the residents.

Paul, or St. Paul as he used to be called on Earth, was the most prolific apostle. He knew Sedona was a wonderful place for people to grow deeply into old souls. He felt blessed to be there and closer to Jesus than he had been in quite some time. God didn't actually tell him Sedona may be a place for a Temple, but Paul thought it sounded like a good idea.

GLOSSARY

Algenia – home planet of Javi and Triol

Bubble – mode of transportation on Lanet

Echoridge – sonar-sensing organ

Far – home planet of Pastor and people that speak in one-word sentences

Ferista race – the Far people

Frond – home planet of Voder and Zac

Ke, kes – genderless pronoun

Lanet – home planet of Aron

Narth – home planet of Dora

Old Souls - spend more time recognizing the soul, find it's the way God wants it to be

Unity of Souls – place where meditation leads the soul, a spiritual place to love others

About The Author

Joan A. Aebig has grown spiritually through participation in many church activities including the Flower Valley Small Group, Rockville, MD, and Appalachia Service Project which she and her husband led at Oakdale Emory United Methodist Church, Olney, MD. To watch teens work and grow spiritually has been one of the joys of her life.

Joan has two adult sons and a daughter-in-law, and enjoys gardening, sewing, and helping deliver dinners to people in need. She has an M.S. in Biomedical Sciences from Hood College, Frederick, MD. She is retired from having spent 33 years in genetic engineering and biomedical research in government and not-for profit labs. Her final position was in NIH's Laboratory of Malaria Immunology and Vaccinology. She is joyfully learning how to write non-science.

Made in the USA
Monee, IL
04 November 2026

67378735R00115